Spring Boot
技术实践

张子宪◎编著

清华大学出版社
北京

内容简介

本书以市面上流行的 Spring Boot 框架开发微服务应用程序为核心，依次引入了 Spring Boot 开发基础，使用 Spring Boot 创建 RESTful API，通过整合 Swagger 2 来自动生成接口文档，使用 Validation 实现对 RESTful 服务的验证，以及使用 Spring Boot 创建前后端分离的应用程序等知识和技术点。

本书共 7 章。第 1 章重点介绍如何使用 Spring Initializr 生成 Spring Boot 项目；第 2 章重点介绍 Spring Boot 快速上手微服务开发；第 3 章重点介绍 Spring Boot 持久性存储的 CRUD 操作及 MyBatis 数据持久化框架；第 4 章重点介绍 OAuth 2 授权框架；第 5 章重点介绍使用 Spring Data Elasticsearch 实现搜索功能；第 6 章重点介绍使用 Spring Boot 创建前后端分离的 Web 应用程序；第 7 章重点介绍 Spring Boot 应用程序监控。

本书适合对开发微服务应用感兴趣的读者阅读和学习，也适合对互联网行业感兴趣的读者选用。

本书封面贴有清华大学出版社防伪标签，无标签者不得销售。

版权所有，侵权必究。举报：010-62782989，beiqinquan@tup.tsinghua.edu.cn。

图书在版编目（CIP）数据

Spring Boot 技术实践 / 张子宪编著. —北京：清华大学出版社，2021.5（2024.8 重印）
ISBN 978-7-302-57732-4

Ⅰ．①S… Ⅱ．①张… Ⅲ．①JAVA 语言—程序设计 Ⅳ．①TP312.8

中国版本图书馆 CIP 数据核字（2021）第 050133 号

责任编辑：张　敏
封面设计：杨玉兰
责任校对：徐俊伟
责任印制：丛怀宇

出版发行：清华大学出版社
网　　址：https://www.tup.com.cn，https://www.wqxuetang.com
地　　址：北京清华大学学研大厦 A 座　　邮　　编：100084
社 总 机：010-83470000　　邮　　购：010-62786544
投稿与读者服务：010-62776969，c-service@tup.tsinghua.edu.cn
质量反馈：010-62772015，zhiliang@tup.tsinghua.edu.cn

印 装 者：北京鑫海金澳胶印有限公司
经　　销：全国新华书店
开　　本：185mm×260mm　　印　张：10　　字　数：247 千字
版　　次：2021 年 6 月第 1 版　　印　次：2024 年 8 月第 4 次印刷
定　　价：49.00 元

产品编号：089800-01

前言

当前，以 Spring Boot 框架为基础的 Java 微服务开发技术逐渐成熟。为了满足广大读者对于使用流行的 Spring Boot 框架开发微服务应用程序的需求，编者积极与合作公司一起探讨，才有了本书的应运而生。

本书共 7 章：

第 1 章：介绍开发微服务应用程序所需的软件环境，特别是 IDEA 集成开发环境，以及如何使用 Spring Initializr 生成 Spring Boot 项目等内容。

第 2 章：介绍测试 RESTful API 的 curl 指令，以及如何使用 Spring Boot 构建 RESTful API 等内容。

第 3 章：介绍 Spring Boot 持久性存储的 CRUD 操作及 MyBatis 数据持久化框架等内容。

第 4 章：介绍 Spring Security 和 Keycloak 服务器实现的 OAuth 2 授权框架等内容。

第 5 章：重点介绍 Spring Boot 整合 Solr 和 Elasticsearch 实现搜索功能等内容。

第 6 章：重点介绍用 Spring Boot 框架实现后端、React 框架实现前端的 Web 应用程序开发等内容。

第 7 章：介绍 Spring Boot Actuator 监控 Spring Boot 应用程序及 Elastic 栈日志监控应用程序等内容。

本书适用于开发微服务应用程序的 Java 开发人员和架构师，对于互联网等相关领域的研究人员和开发微信小程序后端的开发者也有参考价值。广大读者可以通过阅读本书，使用 Spring Boot 框架创建一个微服务架构。

本书由张子宪编著，还获得了早期合著者、合作伙伴、同事、学员、读者、相关合作公司的支持，在此一并表示感谢。

在本书的编写过程中虽力求严谨，但由于时间有限，疏漏之处在所难免，望广大读者批评指正。

编者

目录

第 1 章　Spring Boot 开发基础 1
- 1.1　准备工作环境 1
- 1.2　项目构建工具 2
  - 1.2.1　Maven构建工具 2
  - 1.2.2　Gradle构建工具 4
- 1.3　Spring Initializr 生成 Spring Boot 项目 6
- 1.4　Jenkins 持续集成 7
- 1.5　Linux 操作系统基础 8
  - 1.5.1　SSH远程登录 8
  - 1.5.2　Linux Shell脚本基础 10
  - 1.5.3　Shell脚本基本语法 10
- 1.6　本章小结 13

第 2 章　微服务 14
- 2.1　测试 RESTful API 的 curl 指令 14
- 2.2　JSON 数据格式 15
- 2.3　构建 RESTful API 16
- 2.4　配置文件 20
- 2.5　整合 Swagger 2 22
- 2.6　自定义 Web 控制器参数 24
- 2.7　使用 Validation 实现 RESTful 服务的验证 27
- 2.8　启用 HTTPS 31
- 2.9　本章小结 31

第 3 章　访问数据库 32
- 3.1　Spring Boot CRUD 操作 32
- 3.2　MyBatis 数据持久化框架 36
- 3.3　使用 HikariCP 连接池 40
- 3.4　缓存 41
- 3.5　MongoDB 数据库 43

3.6 本章小结 ·· 53

第 4 章 权限管理 ·· 55
4.1 Security 实现权限控制 ·· 55
4.2 Shiro 实现权限控制 ·· 57
4.3 集成 JWT 身份验证 ·· 66
4.4 OAuth 2 授权框架 ·· 98
    4.4.1 OAuth 2 资源服务器和 Keycloak 服务器 ···································· 98
    4.4.2 Spring Security 和 Keycloak 保护 Spring Boot 应用程序 ·················· 101
4.5 本章小结 ·· 108

第 5 章 Spring Boot 整合搜索引擎 ·· 109
5.1 用于 Solr 的 Spring Data ·· 109
5.2 用于 Elasticsearch 的 Spring Data ·· 111
5.3 实现自动完成 ·· 118
    5.3.1 自动完成服务器端 ·· 118
    5.3.2 自动完成客户端 ·· 120
5.4 界面国际化 ·· 125
5.5 本章小结 ·· 127

第 6 章 Web 应用程序开发 ·· 128
6.1 使用 Bootstrap 实现搜索结果页面 ·· 128
6.2 重试 ·· 129
6.3 整合 Kafka ·· 131
6.4 测试 ·· 135
6.5 React 框架实现前后端分离的 Web 应用程序 ································ 136
6.6 使用 WebSocket 构建交互式 Web 应用程序 ·································· 140
6.7 本章小结 ·· 144

第 7 章 监控 Spring Boot 应用程序 ·· 145
7.1 Spring Boot Actuator ·· 145
7.2 Elastic 栈日志监控 ·· 146
7.3 本章小结 ·· 149

参考文献 ·· 150

第 1 章

Spring Boot 开发基础

Spring Boot 是用于创建微服务的基于 Java 的开源框架。它是由 Pivotal 公司开发的，用于构建独立的和生产就绪的 Spring 应用程序。

Spring Boot 源于 Spring 框架。Spring 框架的第一个版本是由 Rod Johnson 撰写的。他在 2000 年为伦敦的金融界提供独立咨询业务时编写了 Spring 框架最开始的部分。2002 年 10 月 Wrox 公司出版了他的《Java 企业应用设计与开发的专家一对一》一书，发布了该框架。该书出版后，基于读者的要求，源代码在开源使用协议下得以提供。一批自愿拓展 Spring 框架的开发人员组成了团队，2003 年 2 月在 Sourceforge 上构建了一个项目。在 Spring 框架上工作一年后，这支团队在 2004 年 3 月发布了第一个版本（1.0）。继这个版本后，Spring 框架在 Java 社区变得异常流行，部分归结于它作为一个开源项目却有相对较好的文档和参考文献。

为了简化 Spring 的 Web 应用程序架构，2013 年 Spring Boot 诞生。在 Spring Boot 之前的 Spring Web 应用程序需要打包成.war 文件，然后才能部署到 Tomcat 中。Spring Boot 可以使用内嵌的 Tomcat，由 main 方法来启动 Spring，接着启动 Tomcat，而不是由 Tomcat 来启动 Spring。EmbeddedServletContainerAutoConfiguration 会进行 Tomcat 的配置，由 TomcatEmbeddedServletContainer 进行启动。

1.1 准备工作环境

首先要准备一个 Java 开发环境。当前选用 JDK 11（可以从 OpenJDK 的官方网站下载得到），使用默认方式安装即可。然后准备一个用于编写代码的集成开发环境。例如，可以使用 IntelliJ IDEA、Eclipse、Visual Studio Code 或 NetBeans。

如果使用 IDEA 集成开发环境，则为了防止控制台输出的中文显示为乱码，可以通过 Help→Edit Custom VM OPtions 菜单添加-Dfile.encoding = UTF-8 参数，然后重启 IDEA。另外，可以设置项目的默认编码为 UTF-8。

IDEA 的配置信息默认保存在 C 盘，如果需要修改配置信息所在的路径，可以修改文件 idea.properties。具体配置如下：

```
idea.config.path=E:/soft/.IdeaIC/config
idea.system.path=E:/soft/.IdeaIC/system
```

如果使用 NetBeans 集成开发环境，为了让 NetBeans 更好地支持 UTF-8 编码的源文件。可以导航到<Netbeans 安装目录>/etc 并打开 netbeans.conf 文件。在以 netbeans_default_options 开头的行的末尾添加 -J-Dfile.encoding = UTF-8（确保包括前导空格）。重新启动 NetBeans，此时它应该运行在 UTF-8 状态中。若要进行验证，请前往"帮助"→"关于"菜单，并检查系统。

1.2 项目构建工具

关于项目构建工具，本节首先介绍 Maven 构建工具，然后介绍 Gradle 构建工具。

1.2.1 Maven 构建工具

准备 Maven 构建工具，可以下载最新版本的 Maven，当前选用版本是 maven-3.6.3。解压下载的 Maven 压缩文件到 C 盘根目录下，将创建一个 C:\apache-maven-3.6.3 路径。修改 Windows 操作系统环境变量 PATH，增加当前路径 C:\apache-maven-3.6.3\bin。如果一个项目的源代码根路径下包括一个 pom.xml 文件，则说明这个项目可能是用 Maven 构建的。大部分用 Maven 构建的项目只需要执行如下一条命令：

```
#mvn clean install
```

可以在 PowerShell 下使用 scoop 命令安装 Maven：

```
>scoop install maven
```

在项目的根目录中放置 pom.xml，在 src/main/java 目录中放置项目的运行代码，在 src/test/java 目录中放置项目的测试代码。

使用 maven archetype 来创建项目的结构（采用 Maven 构建的项目一般包括一个 pom.xml 文件）。

```xml
<build>
    <plugins>
        <plugin>
            <artifactId>maven-assembly-plugin</artifactId>
            <configuration>
                <archive>
                    <manifest>
                        <mainClass>fully.qualified.MainClass</mainClass>
                    </manifest>
                </archive>
                <descriptorRefs>
                    <descriptorRef>jar-with-dependencies</descriptorRef>
                </descriptorRefs>
            </configuration>
        </plugin>
    </plugins>
</build>
```

使用下面的命令执行它：

```
mvn assembly:single
```

用 install 参数下载依赖的.jar 文件：

```
mvn install
```

Maven 默认的本地仓库地址为${user.home}/.m2/repository。例如，如果用 Administrator 账户登录，则把.jar 文件下载到 C:\Users\Administrator\.m2\repository\这样的路径。

如果.jar 文件位于 lib 路径下，则 Eclipse 的.classpath 文件中的 classpathentry 是 lib 类型的。

```
<classpathentry kind="lib" path="lib/commons-io-1.2.jar"/>
```

如果.jar 文件位于 Maven 的存储库中，则 Eclipse 的.classpath 文件中的 classpathentry 是 var 类型的。

```
<classpathentry kind="var" path="M2_REPO/junit/junit/4.8.2/junit-4.8.2.jar"
             sourcepath="M2_REPO/junit/junit/4.8.2/junit-4.8.2-sources.jar"/>
```

盖大楼时需要搭建最终不会交付使用的脚手架。很多单元测试代码也不会在正式环境中运行，但是必须编写的。与此类似，可以使用 JUnit 做单元测试。maven-surefire-plugin 是 Maven 中执行测试用例的插件。从 Maven 2.22.0 版本开始，Maven Surefire 和 Maven Failsafe 为在 JUnit 平台上执行测试提供本地支持。

POM 文件的构建部分如下所示。

```xml
<build>
    <plugins>
        <plugin>
            <groupId>org.apache.maven.plugins</groupId>
            <artifactId>maven-surefire-plugin</artifactId>
            <version>2.22.1</version>
        </plugin>
    </plugins>
</build>
```

如果使用 Maven Surefire 插件的原生 JUnit 5 支持，必须确保从类路径中找到至少一个测试引擎实现。所以在配置 Maven 构建的依赖项时，需将 junit-jupiter-engine 依赖项添加到测试范围。

```xml
<dependencies>
    <dependency>
        <groupId>org.junit.jupiter</groupId>
        <artifactId>junit-jupiter-api</artifactId>
        <version>5.4.0</version>
        <scope>test</scope>
    </dependency>
    <dependency>
        <groupId>org.junit.jupiter</groupId>
        <artifactId>junit-jupiter-engine</artifactId>
        <version>5.4.0</version>
        <scope>test</scope>
    </dependency>
</dependencies>
```

如果使用 Maven Surefire 插件的默认配置，它会运行从测试类中找到的所有测试方法。测试类的名称具有如下特点：
- 以字符串 Test 开头或结尾。
- 以字符串 Tests 结尾。
- 以字符串 TestCase 结尾。

1.2.2　Gradle 构建工具

相比 Maven，Gradle 构建工具提供了更加灵活的构建方式。

可以下载二进制文件来安装 Gradle：

```
# cd /opt/
# wget -bc https://services.gradle.org/distributions/gradle-5.2-bin.zip
# unzip gradle-5.2-bin.zip
# mv gradle-5.2 ./gradle
```

可以在/etc/profile 文件或者/etc/profile.d 目录设置环境变量。不过/etc/profile.d/比/etc/profile 好维护，对于不需要软件的变量，可以直接删除/etc/profile.d/下对应的 Shell 脚本。

创建设置环境变量的脚本文件：

```
# echo 'export GRADLE_HOME=/opt/gradle' > /etc/profile.d/gradle.sh
# echo 'export PATH=$PATH:$GRADLE_HOME/bin' >> /etc/profile.d/gradle.sh
```

执行以上这个脚本文件：

```
# ./etc/profile.d/gradle.sh
```

检查 gradle 的环境变量是否设置正确：

```
# gradle -v
```

这里，只需要使用 gradle 编译代码：

```
# gradle build
```

如果需要从本地目录中获取.jar 文件，则在 build.gradle 文件中添加如下代码：

```
repositories {
  flatDir {
    dirs 'libs'
  }
}

dependencies {
  compile name: 'commons-exec-1.3'
}
```

如果需要把依赖包下载到本地目录，则可以在 build.gradle 文件中增加如下任务：

```
task copyDependencies(type: Copy) {
  from configurations.compile
  into 'lib'
}
```

build.gradle 中通常包含两个存储库和两个依赖项部分，其中一个存储库完全包含在 buildscript {} 中。buildscript {}部分的依赖项和用于查找它们的存储库仅适用于 build.gradle 脚本本身的代码。这些依赖项通常是 gradle 插件，有时是用于自定义构建代码的依赖项。

例如，一个使用 Spring Boot Gradle 插件的 build.gradle 文件代码如下：

```
buildscript {
    ext {
        springBootVersion = '2.1.2.RELEASE'
    }
    repositories {
        maven {url 'http://maven.aliyun.com/nexus/content/groups/public/'}
        mavenCentral()
    }
    dependencies {
        classpath("org.springframework.boot:spring-boot-gradle-plugin:${springBootVersion}")
    }
}

apply plugin: 'java'
apply plugin: 'eclipse-wtp'
apply plugin: 'org.springframework.boot'
apply plugin: 'io.spring.dependency-management'

group = 'com.lietu'
version = '0.0.1-SNAPSHOT'
sourceCompatibility = '1.8'

repositories {
    maven {url 'http://maven.aliyun.com/nexus/content/groups/public/'}
    mavenCentral()
}

configurations {
    providedRuntime
}

dependencies {
    implementation 'org.springframework.boot:spring-boot-starter-freemarker'
    implementation 'org.springframework.boot:spring-boot-starter-web'
    providedRuntime 'org.springframework.boot:spring-boot-starter-tomcat'
    testImplementation 'org.springframework.boot:spring-boot-starter-test'
    testCompile group: 'junit', name: 'junit', version: '4.11'
}
```

要创建一个 Java 项目，先创建一个新的项目目录，进入并执行：

```
# gradle init --type java-library
```

会得到源文件夹和 Gradle 构建文件。

如果使用默认的 Gradle 包结构创建项目，则有：

```
src/main/java
src/main/resources
src/test/java
```

```
src/test/resources
```

如果需要将已有的 Maven 项目转换为 Gradle 项目，则可以在项目根目录下执行如下命令来生成 build.gradle 文件。

```
#gradle init
```

不需要修改 sourceSets 来执行测试。Gradle 会发现测试类和资源都在 src/test 中，然后可以执行如下命令：

```
#./gradlew test
```

可以编写测试类，然后使用 Gradle 运行一个测试用例：

```
# gradle test -Dtests.class=LibraryTest
# gradle test "-Dtests.class=*.ClassName"
```

运行包和子包中所有的测试用例：

```
# gradle test "-Dtests.class= com.lietu.package.*"
```

Gradle 构建项目的默认行为是执行测试。在构建时，也可以不执行测试，只需执行以下命令：

```
# gradle build -x test
```

大多数工具需要在计算机上进行安装才能使用它们，这可能会给构建用户造成不必要的负担。同样重要的是，用户是否会为构建工具安装正确版本的工具。如果正在使用旧版本的构建软件又该怎么办？Gradle Wrapper（以下称为 Wrapper）解决了这两个问题，是开始 Gradle 构建的首选方式。使用 Wrapper 可以使项目组成员不必预先安装好 Gradle，便于统一项目所使用的 Gradle 版本。如果需要，也可以手工修改 gradle-wrapper.properties 文件中的 distributionUrl 值以改变 Gradle 版本。

在 Windows 操作系统下可以添加环境变量 GRADLE_USER_HOME 自定义 Gradle 缓存位置。

1.3 Spring Initializr 生成 Spring Boot 项目

使用 Spring Initializr 可以基于用户选择的多个依赖项来生成 Spring Boot 项目。

首先访问 Spring Initializr 网址，然后填写项目元数据的详细信息，在 Group 项中输入 com.lietu，在 Artifact 项中输入 SpringInitExample。根据自己的需求选择依赖项，然后单击 GENERATE 按钮生成项目。接下来将会获得一个.zip 解压缩文件，并在 IDE 中打开它。打开项目时，可以自动获得 SpringInitExampleApplication.java 类。SpringInitExampleApplication.java 类是该项目中的主要配置类，具体代码如下：

```
package com.lietu.SpringInitExample;
import org.springframework.boot.SpringApplication;
import org.springframework.boot.autoconfigure.SpringBootApplication;

@SpringBootApplication
public class SpringInitExampleApplication {
    public static void main(String[] args) {
```

```
        SpringApplication.run(SpringInitExampleApplication.class, args);
    }
}
```

控制器控制 HTTP 请求与应用程序逻辑之间的交互。在 com.lietu.SpringInitExample.controller 包中创建控制器类 HelloController，具体代码如下：

```
package com.lietu.SpringInitExample.controller;
import org.springframework.web.bind.annotation.RequestMapping;
import org.springframework.web.bind.annotation.RestController;

@RestController
public class HelloController {
    @RequestMapping("/")
    public String index() {
        return "Hello World Spring !";
    }
}
```

运行 SpringInitExampleApplication 类。打开 Web 浏览器并输入 localhost:8080/，获得输出结果如下所示。

```
Hello World Spring !
```

这时，Spring Boot 使用嵌入的 Tomcat 提供 Web 服务。

1.4 Jenkins 持续集成

为了让产品可以快速迭代，频繁地（一天多次）将代码集成到主干的这种软件开发行为称为持续集成。操作时，可以使用 Jenkins 进行持续集成，并使用 Gradle 或 Maven、MSBuild 等构建工具。

首先从国内的镜像站下载 Jenkins 最新的 war 包，然后安装 Jenkins 插件。

```
#wget https://mirrors.tuna.tsinghua.edu.cn/jenkins/war/latest/jenkins.war
```

Jenkins 的 war 包内部自带了 Web 容器 Jetty。如下命令使用 war 包启动 Jenkins：

```
# java -jar jenkins.war
```

如果要在 Tomcat 上安装 Jenkins，只须将 jenkins.war 文件复制到$TOMCAT_HOME/webapps，然后访问 http://yourhost/jenkins。

登录 Jenkins 管理界面需要密码。当 Tomcat 解压缩 jenkins.war 文件时，会在控制台显示密码。要解锁 Jenkins，请粘贴从文件 C:\Users\<your_user_account_folder_name>\.jenkins\secrets\initialAdminPassword 复制的随机密码。将其粘贴到管理员密码字段中，然后单击"继续"按钮。

Jenkins 将数据存储在 JENKINS_HOME 指定的路径下。JENKINS_HOME 的默认值是 C:\Users\<your_user_account_folder_name>\.jenkins。新建 JENKINS_HOME 系统环境变量，在变量值处输入 Jenkins 安装目录，如"D:\soft\data\jenkins"。也可以在使用 Jenkins 一段时间后更改此

位置。为此，请完全停止 Jenkins，将内容从旧的 JENKINS_HOME 移到新位置，设置新的 JENKINS_HOME，然后重新启动 Jenkins。

下面介绍在 Jenkins 上如何执行 Gradle 应用程序。

在浏览器中，使用端口 8080 导航到 localhost，以呈现 Jenkins 仪表板。系统将要求设置一个新的管理用户及要安装的插件。

在"管理 Jenkins"→"管理插件"下，确保已安装以下两个插件。

- Git plugin。
- Gradle plugin。

接下来，可以设置任务来构建项目。

只须单击几下即可设置新的 Gradle 任务。从左侧导航栏中选择"新建项目"→"自由式项目"。输入项目的名称"gradle-site-plugin"。

在"源代码管理"部分中选中 Git 单选按钮。输入 GitHub 存储库的 URL。

此外，通过选择"Invoke Gradle build script"在"构建"部分中创建构建步骤。我们将要使用包装器 Wrapper 执行构建。在 Tasks 输入框中，输入 build 构建并使用-Switches 选项的--scan -s 生成构建扫描，在构建失败的情况下输出堆栈跟踪。

保存任务的配置并通过触发"立即构建"按钮来执行初始构建。构建应该能成功完成，并呈现一个 Gradle Build Scan 图标。该图标可直接将你带到给定构建的构建扫描（build scan）。

1.5 Linux 操作系统基础

很多 Web 应用后台都运行在 Linux 操作系统中。Linux 来源于 UNIX，是 UNIX 操作系统的开放源代码实现。本节先介绍 SSH 远程登录，然后介绍 Linux Shell 脚本基础及 Shell 脚本基本语法，包括使用 Shell 脚本启动和停止 Spring Boot 应用程序。

1.5.1 SSH 远程登录

通常，我们可以通过 SSH 客户端软件连接到远程的 Linux 服务器。在大多数 Linux 发行版中，SSH 服务器通常作为易于安装的软件包提供给用户，用户可以尝试使用 ssh localhost 命令来测试它是否正在运行。

如果有现成的 Linux 服务器可用，可以使用支持 SSH 协议的终端仿真程序 SecureCRT 连接到远程 Linux 服务器。因为它可以保存登录密码，比较方便使用。除了 SecureCRT，还可以使用开源软件 PuTTY，或者使用可保存登录密码的 KiTTY 及 Xshell。如果使用 root 账户登录，则终端提示符为"#"，否则为"$"。

在 Windows 操作系统下可以安装 Cygwin，使用它来练习 Linux 常用命令。使用 VMware，Linux 可以运行在 Windows 操作系统下。VMware 让 Linux 运行在虚拟机中，而且不会破坏原来的 Windows 操作系统。首先要准备好 VMware，当然仍需要准备 Linux 光盘文件。

就好像华山派有剑宗和气宗，Linux 也有很多种版本，例如 RedHat、Ubuntu 及 SUSE 等。这里介绍 Ubuntu 和 CentOS。

操作系统中可能会安装好几个版本的 JDK。在 Linux 中，为了切换 JDK 版本，只需要修改 /etc/alternatives 中的符号链接指向。

在 Ubuntu 中，如果需要安装软件，可以下载 deb 安装包，然后使用 dpkg 命令安装。但一个软件包可能依赖其他的软件包。因此，为了安装一款软件，可能需要下载其他的好几个它所依赖的软件包。为了简化安装操作，可以使用高级包装工具（Advanced Packaging Tool，APT）。APT 会自动计算出程序之间的相互关联性，并且计算出完成软件包的安装需要哪些步骤。这样，在安装软件时，不会再被那些关联性问题所困扰。

在/etc/apt/sources.list 文件中指示了包的来源的存储库。包的来源可以是 CD 或 DVD，硬盘上的目录或 HTTP/FTP 服务器上的目录。请求的数据包位于服务器（或本地硬盘）上，它将自动下载并安装。APT 主要关注采购包、包的可用版本的比较及包文件的管理。实际上，可以通过浏览器浏览在 FTP 或 HTTP 上的存储库。

如果需要修改/etc/apt/sources.list 文件，可以执行如下命令先备份这个文件。

```
# sudo cp /etc/apt/sources.list /etc/apt/sources.list.bak
```

如果这一步出现"sudo: unable to resolve host t-000004"这样的错误，则可以考虑执行如下命令修改/etc/hosts 文件的内容。

```
# echo $(hostname -I | cut -d\-f1) $(hostname) | sudo tee -a /etc/hosts
```

如果安装过程中出现"E: Could not get lock /var/lib/dpkg/lock"这样的错误，则可以尝试使用如下命令修复。

```
# sudo fuser -cuk /var/lib/dpkg/lock
# sudo rm -f /var/lib/dpkg/lock
```

在 CentOS 中，如果需要安装软件，可以下载 RPM 安装包，然后使用 RPM 安装。例如，下载 Elasticsearch 软件的安装包 elasticsearch-6.6.0.rpm：

```
# wget https://artifacts.elastic.co/downloads/elasticsearch/elasticsearch-6.6.0.rpm
```

使用如下命令安装：

```
# rpm -ivh elasticsearch-6.6.0.rpm
```

但有些操作系统对应的 RPM 安装包找起来往往比较麻烦。为了简化安装操作，可以使用《黄狗升级管理器》（Yellow dog Updater Modified），一般简称 yum。

yum 软件包管理器自动从网络下载并安装软件。yum 有点类似《360 软件管家》，但是不会有商业倾向的推销软件。例如，安装支持 wget 命令的软件：

```
#yum install wget
```

为了方便在服务器端编写 Shell 脚本，可以采用 Micro 这样的终端文本编辑器。

在 Linux 上，可以通过 snap 命令安装 Micro：

```
# snap install micro --classic
```

保存文件后，按 Ctrl+Q 组合键退出。

1.5.2 Linux Shell 脚本基础

Shell 是一款命令语言解释器，拥有内置的 Shell 命令集。此外，Shell 也能被系统中其他有效的 Linux 实用程序和应用程序所调用。Shell 是用户与 Linux 内核之间的接口程序。用户在命令行提示符下输入的每条命令都由 Shell 先解释，再传给 Linux 内核。

Shell 具备的主要功能如下。

- 命令解释：将用户可读的命令转换成计算机可理解的命令，并控制命令执行。
- 输入/输出重定向：操作系统将键盘作为标准输入、显示器作为标准输出。当这些定向不能满足用户需求时，用户可以在命令中用符号 ">" 或 "<" 重新定向。
- 管道处理：利用管道将一个命令的输出送入另一个命令，实现多个命令组合完成复杂命令的功能。
- 系统环境设置：用 Shell 命令设置系统环境变量，维护用户的工作环境。
- 程序设计语言：Shell 命令本身可以作为程序设计语言，将多个 Shell 命令组合起来，编写能实现系统或用户所需功能的程序。

Linux 提供了很多种 Shell，例如 Z shell、Pash 和 Fish 等，但一般使用 Bash 脚本。

1.5.3 Shell 脚本基本语法

在屏幕上打印 "Hello"：
```
echo "Hello"
```
将 ABC 分配给 a：
```
a=ABC
```
输出 a 的值：
```
echo $a
```
在屏幕上打印 "ABC"。

将 ABC.log 分配给 b：
```
b=$a.log
```
输出 b 的值：
```
# echo $b
ABC.log
```
把文件 ABC.log 的内容写入 testfile：
```
cat $b > testfile
```

"指令--help" 会输出帮助信息。

可以把重复执行的 Shell 脚本写入一个文本文件。在 Linux 中，文件扩展名不作为系统识别文件类型的依据，但是可以作为识别文件的依据，例如可以简单地将脚本文件以.sh 结尾。

在 Linux 下，可以通过 vi 命令创建一个诸如 script.sh 的文件：vi script.sh。创建好脚本文件后，就可以在文件内用脚本语言要求的格式编写脚本程序。

在创建的脚本文件中输入以下代码，并保存和退出。

```
#! /bin/bash
echo "Hello world!"
```

添加脚本文件的可执行运行权限 chmod 777 script.sh，运行文件./script.sh，得到结果如下：

```
Hello world!
```

Shell 脚本中用#表示注释，相当于 C 语言中的注释符//。但如果#位于第一行开头，并且是#!（称为 Shebang）则例外，它表示该脚本使用后面指定的解释器/bin/sh 解释执行。每个脚本程序必须在开头包含这个语句。

使用参数-n 检查语法错误，例如：

```
# bash -n ./test.sh
```

如果 Shell 脚本有语法错误，则会提示错误所在行；否则，不输出任何信息。

if 语句的语法格式如下：

```
if [ condition ] then
    command1
elif    # elif 与 else if 等价
    then
        command2
    else
        default-command
fi  # 这里的 fi 就是 if 反过来写的形式
```

例如，为了判断某个命令是否存在，可以使用如下代码。

```
if which programname >/dev/null; then
    echo exists
else
    echo does not exist
fi
```

再如，判断 yum 是否存在，可以使用如下代码。

```
if which yum >/dev/null; then
    echo "exists"
else
    echo "does not exist"
fi
```

case 语句的语法格式如下：

```
case 字符串 in
    模式1)
        语句
        ;;
    模式2)
        语句
        ;;
    *)
        默认执行的语句
        ;;
esac
```

这里的 esac 就是 case 反过来写的形式。例如：

```
extension="png"
```

```
case "$extension" in
  "jpg"|"jpeg")
    echo "It's image with jpeg extension."
  ;;
  "png")
    echo "It's image with png extension."
  ;;
  "gif")
    echo "Oh, it's a giphy!"
  ;;
  *)
    echo "Woops! It's not image!"
  ;;
esac
```

在以上代码中，使用"|"将"jpg"和"jpeg"这两个模式连接到了一起。

下面介绍 4 种模式匹配形式。

- ${variable#pattern}：从$string 的前面删除$substring 的最短匹配。
- ${variable##pattern}：从$string 的前面删除$substring 的最长匹配。
- ${variable%pattern}：从$string 的后面删除$substring 的最短匹配。
- ${variable%%pattern}：从$string 的后面删除$substring 的最长匹配。

应用模式匹配的例子如下：

```
x=/home/cam/book/long.file.name
echo ${x#/*/}
echo ${x##/*/}
echo ${x%.*}
echo ${x%%.*}
```

输出结果如下：

```
cam/book/long.file.name
long.file.name
/home/cam/book/long.file
/home/cam/book/long
```

安装 Fish Shell：

```
# sudo apt-get install fish
```

启动 Spring Boot Web 应用程序的脚本 startup.sh 内容如下。

```
#!/bin/bash
nohup java -jar /path/to/app/hello-world.jar > /path/to/log.txt 2>&1 &
echo $! > /path/to/app/pid.file
```

其中，nohup 允许 Java 进程即使在用户注销后仍在后台运行；这里的 echo $! 将打印最后一条命令的 PID，然后使用> /path/to/app/pid.file 将 PID 写入文件。

停止应用程序的脚本 shutdown.sh 内容如下。

```
#!/bin/bash
kill $(cat /path/to/app/pid.file)
```

shutdown.sh 脚本使用了由 startup.sh 脚本创建的 PID 文件，这使我们能够找到并"杀死"以前产生的确切进程。

1.6　本章小结

本章首先介绍了开发 Spring Boot 的环境，然后介绍了自动化构建工具 Maven 和 Gradle，以及使用 Spring Initializr 生成 Spring Boot 项目，最后介绍了持续集成软件 Jenkins 以及使用 Jenkins 构建应用程序等内容。

第 2 章

微服务

本章首先介绍通过 curl 指令来测试 RESTful API；Spring Boot 默认使用 JSON 作为响应报文格式，因此随后介绍 JSON 数据格式；使用 Spring Boot 可以创建 Web 服务，并且可以使用 JavaScript 框架将其在 Web 前端展示，因此还会介绍使用 RESTful API 实现 Web 服务等内容。

2.1 测试 RESTful API 的 curl 指令

使用 curl 指令可以发送 HTTP 网络请求来测试 RESTful API。API 请求由以下 4 个不同部分组成。

- endpoint：这是客户端用于与服务器通信的 URL。
- HTTP 方法：用来告知服务器端和客户端要执行什么操作。最常见的方法是 GET、POST、PUT、DELETE 和 PATCH。
- 标头：用来在服务器端和客户端之间传递其他信息，例如授权。
- body：发送到服务器的数据。

curl 命令的语法格式如下：

```
curl [options] [URL…]
```

在发出请求时，可以使用以下选项。

- -X, --request：表示使用的 HTTP 方法。
- -i, --include：表示包括响应头。
- -d, --data：表示要发送的数据。
- -H, --header：表示要发送的其他标头。

下面结合 curl 命令对上述 GET、POST、PUT 等方法进行介绍。

（1）GET 方法用来从服务器请求特定资源。使用 curl 指令发出 HTTP 请求时，GET 方法是默认方法。以下命令实现向 JSONPlaceholder API 发出 GET 请求。

```
# curl https://jsonplaceholder.typicode.com/posts
```

若使用查询参数过滤结果，则使用以下命令。

```
# curl https://jsonplaceholder.typicode.com/posts?userId=1
```

（2）POST 方法用来在服务器上创建资源。如果资源存在，则将其覆盖。以下命令将使用-d 选项指定的数据创建一个新帖子。

```
# curl -X POST -d "userId=5&title=Hello World&body=Post body." https://jsonplaceholder.typicode.com/posts
```

请求主体的类型使用 Content-Type 标头指定。默认情况下，如果不指定此标头，curl 指令将使用 Content-Type: application/x-www-form-urlencoded 形式。要发送 JSON 格式的数据，则将主体类型设置为 application/json。例如：

```
#curl -X POST -H "Content-Type: application/json" \
    -d '{"userId": 5, "title": "Hello World", "body": "Post body."}' \
    https://jsonplaceholder.typicode.com/posts
```

（3）PUT 方法用于更新或替换服务器上的资源，将指定资源的所有数据替换为请求数据。例如：

```
#curl -X PUT -d "userId=5&title=Hello World&body=Post body." https://jsonplaceholder.typicode.com/posts/5
```

（4）PATCH 方法用于对服务器上的资源进行部分更新。例如：

```
#curl -X PATCH -d "title=Hello Universe" https://jsonplaceholder.typicode.com/posts/5
```

（5）DELETE 方法用于从服务器中删除指定的资源。例如：

```
#curl -X DELETE https://jsonplaceholder.typicode.com/posts/5
```

另外，如果 API 端点需要身份验证，则需要获取访问密钥；否则，API 服务器将使用"禁止访问"或"未经授权"等响应消息来响应。获取访问密钥的过程取决于所使用的 API。获得访问令牌后，可以执行以下指令在标头中发送它。

```
#curl -H "Authorization: Basic <ACCESS_TOKEN>" http://www.example.com
```

2.2 JSON 数据格式

RESTful API 通常以 JSON 数据格式响应。JSON（JavaScript Object Notation）是一种轻量级的数据交换格式，人们很容易阅读和编写它，机器也很容易解析和生成相应文件。可以用它传输由名称-值（键-值）对和数组数据类型组成的数据对象。一些结构复杂的数据也可以采用 JSON 格式表示，例如，哈希表中的值是数组形式的。

JSON 的基本数据类型如下。
- 数字：有符号的十进制数字，可能包含小数部分，可使用指数 E 表示法，但不能包括非数字，如 NaN。该类型不区分整数和浮点数。
- 字符串：0 个或多个 Unicode 字符的序列。字符串用双引号分隔，并支持反斜杠转义语法。
- 布尔值：为 true 或 false 中的任意值。
- 数组：0 个或多个值的有序列表，每个值可以是任意类型。数组使用方括号括起，元素以逗号分隔。
- 对象：名称-值对的无序集合，其中"名称"（也称为"键"）是字符串形式的。由于对象旨在表示关联数组，因此推荐每个键在对象内是唯一的。对象用大括号分隔，并使用逗号分隔每对，而在每一对中用冒号将键或名称与其值分隔开。

- null：一个空值，使用单词 null 表示。

Spring Boot 内置了软件包 Jackson 来完成 JSON 的序列化和反序列化操作。

2.3 构建 RESTful API

举例说明如何构建 RESTful API。首先使用 Spring Initializr 创建一个名为 spring-boot-rest-2 的项目。生成的 SpringBootRest2Application.java 代码如下：

```
package com.bhaiti.kela.server.main;
import org.springframework.boot.SpringApplication;
import org.springframework.boot.autoconfigure.SpringBootApplication;
@SpringBootApplication
public class SpringBootRest2Application {
    public static void main(String[] args) {
        SpringApplication.run(SpringBootRest2Application.class, args);
    }
}
```

Java 的注解是附加在代码中的一些元信息。而 Spring Boot 的注解@SpringBootApplication 等效于使用@Configuration、@EnableAutoConfiguration 和@ComponentScan，并且这些注释经常一起使用。在大多数情况下，Spring Boot 开发中总是使用这 3 个重要的注解对主类进行注解。在这里，我们将使用组件路径来修改@SpringBootApplication；否则，应用程序将无法找到控制器类。

```
@SpringBootApplication(scanBasePackages = {"com.bhaiti"})
public class SpringBootRest2Application {
    public static void main(String[] args) {
        SpringApplication.run(SpringBootRest2Application.class, args);
    }
}
```

创建实体类 Student，代码如下：

```
package com.bhaiti.kela.beans;
public class Student {
    String name;
    int age;
    String registrationNumber;
    public String getName() {
        return name;
    }
    public void setName(String name) {
        this.name = name;
    }
    public int getAge() {
        return age;
    }
    public void setAge(int age) {
        this.age = age;
    }
```

```java
    public String getRegistrationNumber() {
        return registrationNumber;
    }
    public void setRegistrationNumber(String registrationNumber) {
        this.registrationNumber = registrationNumber;
    }
}
```

创建类 StudentRegistration,代码如下:

```java
package com.bhaiti.kela.beans;
import java.util.ArrayList;
import java.util.List;

public class StudentRegistration {
    private List<Student> studentRecords;
    private static StudentRegistration stdregd = null;
    private StudentRegistration() {
        studentRecords = new ArrayList<Student>();
    }

    public static StudentRegistration getInstance() {
        if (stdregd == null) {
            stdregd = new StudentRegistration();
            return stdregd;
        } else {
            return stdregd;
        }
    }

    public void add(Student std) {
        studentRecords.add(std);
    }

    public String updateStudent(Student std) {
        for (int i = 0; i < studentRecords.size(); i++) {
            Student stdn = studentRecords.get(i);
            System.out.println(stdn.getRegistrationNumber());
            System.out.println(std.getRegistrationNumber());
            if (stdn.getRegistrationNumber().equals(std.getRegistrationNumber())) {
                studentRecords.set(i, std);        //update the new record
                return "Update successful";
            }
        }

        return "Update un-successful";
    }

    public String deleteStudent(String registrationNumber) {
        for (int i = 0; i < studentRecords.size(); i++) {
            Student stdn = studentRecords.get(i);
            if (stdn.getRegistrationNumber().equals(registrationNumber)) {
```

```
                studentRecords.remove(i);        //delete the new record
                return "Delete successful";
            }
        }
        return "Delete un-successful";
    }

    public List<Student> getStudentRecords() {
        return studentRecords;
    }
}
```

添加一个名为 StudentRegistrationReply 的类，并进行如下修改。此类用于将响应回复给客户端应用程序。

```
package com.bhaiti.kela.beans;
public class StudentRegistrationReply {
    String name;
    int age;
    String registrationNumber;
    String registrationStatus;
    public String getName() {
        return name;
    }
    public void setName(String name) {
        this.name = name;
    }
    public int getAge() {
        return age;
    }
    public void setAge(int age) {
        this.age = age;
    }
    public String getRegistrationNumber() {
        return registrationNumber;
    }
    public void setRegistrationNumber(String registrationNumber) {
        this.registrationNumber = registrationNumber;
    }
    public String getRegistrationStatus() {
        return registrationStatus;
    }
    public void setRegistrationStatus(String registrationStatus) {
        this.registrationStatus = registrationStatus;
    }
}
```

现在，将介绍两个控制器：一个用于服务 GET 请求；另一个用于服务 POST 请求。通过 GET 请求，可以实现检索所有学生的注册信息，而通过 POST 请求，可以实现将学生信息添加到应用程序中。

下面的 StudentRetrieveController 类将通过返回 JSON 格式的 Student 类对象列表来处理/

student/allstudent 的 GET 请求。

```java
package com.bhaiti.kela.controllers;
import java.util.List;
import org.springframework.stereotype.Controller;
import org.springframework.web.bind.annotation.RequestMapping;
import org.springframework.web.bind.annotation.RequestMethod;
import org.springframework.web.bind.annotation.ResponseBody;
import com.bhaiti.kela.beans.Student;
import com.bhaiti.kela.beans.StudentRegistration;
@Controller
public class StudentRetrieveController {
    @RequestMapping(method = RequestMethod.GET, value="/student/allstudent")
    @ResponseBody
    public List<Student> getAllStudents() {
        return StudentRegistration.getInstance().getStudentRecords();
    }
}
```

在 com.bhaiti.kela.controllers 包中添加控制器类 StudentRegistrationController，代码如下：

```java
package com.bhaiti.kela.controllers;
import java.util.List;
import org.springframework.stereotype.Controller;
import org.springframework.web.bind.annotation.RequestBody;
import org.springframework.web.bind.annotation.RequestMapping;
import org.springframework.web.bind.annotation.RequestMethod;
import org.springframework.web.bind.annotation.ResponseBody;
import com.bhaiti.kela.beans.*;

@Controller
public class StudentRegistrationController {
    @RequestMapping(method = RequestMethod.POST, value="/register/student")

    @ResponseBody
    StudentRegistrationReply registerStudent(@RequestBody Student student) {

        System.out.println("In registerStudent");
        StudentRegistrationReply stdregreply = new StudentRegistrationReply();
        StudentRegistration.getInstance().add(student);
        //We are setting the below value just to reply a message back to the caller
        stdregreply.setName(student.getName());
        stdregreply.setAge(student.getAge());
        stdregreply.setRegistrationNumber(student.getRegistrationNumber());
        stdregreply.setRegistrationStatus("Successful");
        return stdregreply;
    }
}
```

引入控制器类来处理 PUT 请求和 DELETE 请求。创建 StudentUpdateController 类，并进行如下修改。

```java
package com.bhaiti.kela.controllers;
import org.springframework.stereotype.Controller;
```

```java
import org.springframework.web.bind.annotation.RequestBody;
import org.springframework.web.bind.annotation.RequestMapping;
import org.springframework.web.bind.annotation.RequestMethod;
import org.springframework.web.bind.annotation.ResponseBody;
import com.bhaiti.kela.beans.Student;
import com.bhaiti.kela.beans.StudentRegistration;

@Controller
public class StudentUpdateController {
    @RequestMapping(method = RequestMethod.PUT, value = "/update/student")
    @ResponseBody
    public String updateStudentRecord(@RequestBody Student stdn) {
        System.out.println("In updateStudentRecord");
        return StudentRegistration.getInstance().updateStudent(stdn);
    }
}
```

创建 StudentDeleteController 类,并按如下所示进行修改。

```java
package com.bhaiti.kela.controllers;
import org.springframework.stereotype.Controller;
import org.springframework.web.bind.annotation.RequestMapping;
import org.springframework.web.bind.annotation.RequestMethod;
import org.springframework.web.bind.annotation.ResponseBody;
import org.springframework.web.bind.annotation.PathVariable;
import com.bhaiti.kela.beans.StudentRegistration;

@Controller
public class StudentDeleteController {
    @RequestMapping(method = RequestMethod.DELETE, value="/delete/student/{regdNum}")
    @ResponseBody
    public String deleteStudentRecord(@PathVariable("regdNum") String regdNum) {
        System.out.println("In deleteStudentRecord");
        return StudentRegistration.getInstance().deleteStudent(regdNum);
    }
}
```

@RequestBody 会将方法的参数绑定到 HTTP 请求的正文,而@ResponseBody 会对响应和返回类型执行相同的操作。Spring Boot 的@RestController 注释基本上是一个快捷方式,可以帮助我们避免必须定义@ResponseBody 的情况。@RestController 注解是@Controller 和@ResponseBody 注解的组合。

运行服务器后,会执行 POST 调用,将 4 条记录插入,并执行 GET 调用检索记录。

2.4 配置文件

在 Spring Boot 项目中有 3 个属性文件,即 application-dev.properties、application-prod.properties 和 application.properties。可以在 application-dev.properties 文件中指定所有开发属性,并在 application-prod.properties 文件中指定生产配置属性,还可以在 application.properties 文件中指定

激活 dew 配置文件，代码如下。

```
spring.profiles.active=dev
```

或者可以在命令行中使用-Dprofile = 参数来选择/覆盖配置文件。

Spring Boot 支持基于 YAML 的属性配置来运行应用程序，可以使用 application.yml 文件代替 application.properties 文件。该 YAML 文件也应位于类路径中。下面给出了关于 application.yml 文件的示例代码。

```
spring:
  application:
    name: demoservice
  server:
port: 9090
```

@Value 注解用于读取 Java 代码中的环境或应用程序的属性值。其读取属性值的语法如下所示。

```
@Value("${property_key_name}")
```

请看以下示例，该示例显示了使用@Value 注解读取 spring.application.name 属性值的语法。

```
@Value("${spring.application.name}")
```

观察下面给出的代码，以便更好地理解@Value 注解。

```
import org.springframework.beans.factory.annotation.Value;
import org.springframework.boot.SpringApplication;
import org.springframework.boot.autoconfigure.SpringBootApplication;
import org.springframework.web.bind.annotation.RequestMapping;
import org.springframework.web.bind.annotation.RestController;

@SpringBootApplication
@RestController
public class DemoApplication {
  @Value("${spring.application.name}")
  private String name;
  public static void main(String[] args) {
    SpringApplication.run(DemoApplication.class, args);
  }
  @RequestMapping(value = "/")
  public String name() {
    return name;
  }
}
```

注意：如果在运行应用程序时未找到该属性，则 Spring Boot 会抛出"非法参数"异常（因为无法解析值"${spring.application.name}"中的占位符"spring.application.name"）。

要解决占位符问题，我们可以使用下面给出的代码，为属性设置默认值。

```
@Value("${property_key_name:default_value}")
@Value("${spring.application.name:demoservice}")
```

2.5 整合 Swagger 2

Swagger 2 是一个开源项目，用于生成 RESTful Web 服务的 REST API 文档。它提供了一个用户界面，用于通过 Web 浏览器访问 RESTful Web 服务。为了将 Swagger 2 添加到 Spring Boot REST API 项目，需要把 Swagger 2 相关的依赖项添加到 pom.xml。

```xml
<dependency>
    <groupId>io.springfox</groupId>
    <artifactId>springfox-swagger2</artifactId>
    <version>2.9.2</version>
</dependency>
<dependency>
    <groupId>io.springfox</groupId>
    <artifactId>springfox-core</artifactId>
    <version>2.9.2</version>
</dependency>
<dependency>
    <groupId>io.springfox</groupId>
    <artifactId>springfox-swagger-ui</artifactId>
    <version>2.9.2</version>
</dependency>
```

对于 Gradle 用户，在 build.gradle 文件中添加以下依赖项。

```
compile group: 'io.springfox', name: 'springfox-swagger2', version: '2.9.2'
compile group: 'io.springfox', name: 'springfox-core', version: '2.9.2'
compile group: 'io.springfox', name: 'springfox-swagger-ui', version: '2.9.2'
```

下一步是创建配置文件。此配置文件的创建目的是配置项目的 basePackage 和选择器，并使配置的 Docket Bean 在我们的应用程序中可用。创建此配置文件后，幕后框架将完成很多工作。

下面是一个非常简单的配置文件示例。

```java
package com.appsdeveloperblog.app.ws;
import org.springframework.context.annotation.Bean;
import org.springframework.context.annotation.Configuration;
import springfox.documentation.builders.PathSelectors;
import springfox.documentation.builders.RequestHandlerSelectors;
import springfox.documentation.spi.DocumentationType;
import springfox.documentation.spring.web.plugins.Docket;
import springfox.documentation.swagger2.annotations.EnableSwagger2;
@Configuration
@EnableSwagger2
public class SwaggerConfig {
    @Bean
    public Docket apiDocket() {
        Docket docket = new Docket(DocumentationType.SWAGGER_2)
          .select()
          .apis(RequestHandlerSelectors.basePackage("com.appsdeveloperblog.app.ws"))
          .paths(PathSelectors.any())
          .build();
        return docket;
```

            }
    }

上面的代码片段中有以下几个重要的细节需要注意。

- apis()：在这里指定需要包含在 Swagger 中的类。例如，提供了项目的基本程序包，因此将寻找所需类的工作委托给了框架。该基本程序包将由框架进行扫描，并根据它们具有的注释囊括所需的类。
- path()：在这里允许从应该包含的 Controller 类中指定方法（使用@Path 注解进行注释）。由于希望所有方法都包含在使用 Swagger 创建的文档中，因此这里提供了 PathSelectors.any()。

如果 RESTful Web 服务应用程序正在使用 Spring Security，那么将需要在 Java 类中进行一些配置，以扩展 WebSecurityConfigurerAdapter 并使用@EnableWebSecurity 注解进行标记。

在 Spring Boot 项目中打开扩展了 WebSecurityConfigurerAdapter 的 Java 类，并使用@EnableWebSecurity 注解进行标记，然后启用以下路径。

```
.antMatchers("/h2-console/**")
.permitAll()
.antMatchers("/v2/api-docs", "/configuration/**", "/swagger*/**", "/webjars/**")
.permitAll()
.anyRequest().authenticated()
```

下面是配置了 Swagger 和其他端点的 Web 安全类示例。

```java
package com.appsdeveloperblog.app.ws.security;
import com.appsdeveloperblog.app.ws.service.UserService;
import org.springframework.http.HttpMethod;
import org.springframework.security.config.annotation.authentication.builders.AuthenticationManagerBuilder;
import org.springframework.security.config.annotation.web.builders.HttpSecurity;
import org.springframework.security.config.annotation.web.configuration.EnableWebSecurity;
import org.springframework.security.config.annotation.web.configuration.WebSecurityConfigurerAdapter;
import org.springframework.security.config.http.SessionCreationPolicy;
import org.springframework.security.crypto.bcrypt.BCryptPasswordEncoder;
@EnableWebSecurity
public class WebSecurity extends WebSecurityConfigurerAdapter {
    private final UserService userDetailsService;
    private final BCryptPasswordEncoder bCryptPasswordEncoder;
    public WebSecurity(UserService userDetailsService, BCryptPasswordEncoder bCryptPasswordEncoder) {
        this.userDetailsService = userDetailsService;
        this.bCryptPasswordEncoder = bCryptPasswordEncoder;
    }
    @Override
    protected void configure(HttpSecurity http) throws Exception {
        http.csrf().disable().
                authorizeRequests()
```

```
                .antMatchers(HttpMethod.POST, SecurityConstants.SIGN_UP_URL)
                .permitAll()
                .antMatchers("/v2/api-docs", "/configuration/**", "/swagger*/**",
"/webjars/**")
                .permitAll()
                .anyRequest().authenticated().and()
                .addFilter( new AuthenticationFilter(authenticationManager()) )
                .addFilter( new AuthorizationFilter( authenticationManager() ))
                .sessionManagement()
                .sessionCreationPolicy(SessionCreationPolicy.STATELESS);
        http.headers().frameOptions().disable();
    }
    @Override
    public void configure(AuthenticationManagerBuilder auth) throws Exception {
        auth.userDetailsService(userDetailsService).passwordEncoder(bCryptPassword
Encoder);
    }

}
```

如果现在运行应用程序假设正在本地运行 Spring Boot 应用程序，则可以通过在浏览器窗口中输入 URL 地址（http://localhost:8080/<CONTEXT PATH HERE>/v2/api-docs）来获取 RESTful API 的 Swagger JSON 文件生成的内容。

注意：如果在 application.properties 文件中提供了一个端口号，则需要在上面的 URL 中确保提供正确的端口号和上下文路径；如果没有在 application.properties 文件中配置上下文路径，则不需要在上面的 URL 中提供任何上下文路径。

例如，如果应用程序属性文件具有以下设置：

```
server.servlet.context-path=/mobile-app-ws
server.port=8888
```

那么将需要通过以下方式打开/v2/api-docs（http://localhost:8888/mobile-app-ws/v2/api-docs）或 swagger-ui.html（http://localhost:8888/mobile-app-ws/swagger-ui.html）页面。

2.6 自定义 Web 控制器参数

假设正在构建一个管理 GitHub 的 Git 存储库的应用程序。为了标识某个 GitRepository 实体，使用 GitRepositoryId 值对象，而不是使用简单的 Long 值。这样，就不会将存储库 ID 与用户 ID 混淆。现在，想在 Web 控制器的方法签名中使用 GitRepositoryId 而不是 Long，我们不必自己进行这种转换。另一个用例是当想从所有控制器的 URL 路径中提取一些上下文对象时。例如，考虑一下 GitHub 上的存储库名称，每个 URL 都以存储库名称开头，若每次在 URL 中都有一个存储库名称时，希望 Spring Boot 自动将该存储库名称转换为 GitRepository 实体，并将其传递到控制器中以进行进一步处理。再如，希望 Spring 自动将路径变量转换为 GitRepositoryId 对象，代码如下：

```
@RestController
class GitRepositoryController {
  @GetMapping("/repositories/{repoId}")
  String getSomething(@PathVariable("repoId") GitRepositoryId repositoryId) {
  …// load and return repository
  }

}
```

GitRepositoryId 是一个简单的值对象，代码如下：

```
@Value
class GitRepositoryId {
   private final long value;
}
```

这里使用 Lombok 提供的注解@Value，因此不必自行创建构造函数和 get 方法。

实现自定义转换器的代码如下：

```
@Component
class GitRepositoryIdConverter implements Converter<String, GitRepositoryId> {
   @Override
   public GitRepositoryId convert(String source) {
       return new GitRepositoryId(Long.parseLong(source));
   }
}
```

由于来自 HTTP 请求的所有输入均被视为字符串，因此需要构建一个 Converter，将 String 值转换为 GitRepositoryId。通过添加@Component 注解，可以让 Spring Boot 知道此转换器。然后，Spring Boot 将自动将此转换器应用于 GitRepositoryId 类型的所有控制器方法参数。

除了构建转换器，还可以在 GitRepositoryId 值对象上提供一个静态的 valueOf()方法，代码如下：

```
@Value
class GitRepositoryId {
   private final long value;
   public static GitRepositoryId valueOf(String value){
       return new GitRepositoryId(Long.parseLong(value));
   }
}
```

实际上，此方法的作用与我们上面构建的转换器相同（将 String 转换为 value 对象）。如果像这样的方法在用作控制器方法中的参数的对象上可用，则 Spring 将自动调用它以进行转换，而无需单独的 Converter Bean。前面的 Converter 解决方案能够生效是因为我们使用了 Spring 的@PathVariable 注解，将方法参数绑定到 URL 路径中的变量。

现在，假设所有的 URL 都以 Git 存储库的名称开头（称为 URL 友好的"slug"），并且希望最小化样板代码：

- 不希望使用许多@PathVariable 注解来"污染"我们的代码；
- 不希望每个控制器都必须检查 URL 中的存储库别名（slug）是否有效；
- 不希望每个控制器都必须从数据库加载存储库数据。

此时，可以通过构建自定义 HandlerMethodArgumentResolver 来实现。

从所期望的控制器代码的外观开始介绍。

```
@RestController
@RequestMapping(path = "/{repositorySlug}")
class GitRepositoryController {
   @GetMapping("/contributors")
   String listContributors(GitRepository repository) {
     …// list the contributors of the GitRepository
   }
   … // more controller methods
}
```

在类级别的@RequestMapping 注解中，定义所有请求均以{repositorySlug}变量开头。当有人获取/访问路径/{repositorySlug}/contributors/时，将调用 listContributors()方法。该方法需要一个 GitRepository 对象作为参数，以便它知道要使用哪个 Git 存储库。

实现一个自定义的 HandlerMethodArgumentResolver，代码如下：

```
@RequiredArgsConstructor
class GitRepositoryArgumentResolver implements HandlerMethodArgumentResolver {
   private final GitRepositoryFinder repositoryFinder;
   @Override
   public boolean supportsParameter(MethodParameter parameter) {
      return parameter.getParameter().getType() == GitRepository.class;
   }

   @Override
   public Object resolveArgument(
     MethodParameter parameter,
     ModelAndViewContainer mavContainer,
     NativeWebRequest webRequest,
     WebDataBinderFactory binderFactory) {

       String requestPath = ((ServletWebRequest) webRequest)
       .getRequest()
       .getPathInfo();

       String slug = requestPath
       .substring(0, requestPath.indexOf("/", 1))
       .replaceAll("^/", "");

       return GitRepositoryFinder.findBySlug(slug)
       .orElseThrow(NotFoundException::new);
   }
}
```

在以上方法 resolveArgument()中，提取请求路径的第一段，其中应该包含存储库别名。然后，将此别名添加到 GitRepositoryFinder 中，以从数据库加载存储库。如果 GitRepositoryFinder 找不到带有该别名的存储库，将抛出一个自定义的 NotFoundException；否则，返回在数据库中找到的 GitRepository 对象。

接下来，必须让 Spring Boot 知道 GitRepositoryArgumentResolver 的存在，代码如下：

```
@Component
@RequiredArgsConstructor
class GitRepositoryArgumentResolverConfiguration implements WebMvcConfigurer {

    private final GitRepositoryFinder repositoryFinder;

    @Override
    public void addArgumentResolvers(
        List<HandlerMethodArgumentResolver> resolvers) {
        resolvers.add(new GitRepositoryArgumentResolver(repositoryFinder));
    }

}
```

在以上代码中，编程实现了 WebMvcConfigurer 接口，并将 GitRepositoryArgumentResolver 添加到了解析器列表中。另外，不要忘记添加@Component 注解，以使 Spring Boot 知道此配置程序。

最后，将自定义的 NotFoundException 映射到 HTTP 状态代码 404。我们可以通过创建@ControllerAdvice 来做到这一点，代码如下：

```
@ControllerAdvice
class ErrorHandler {
    @ExceptionHandler(NotFoundException.class)
    ResponseEntity<?> handleHttpStatusCodeException(NotFoundException e) {
        return ResponseEntity.status(e.getStatusCode()).build();
    }

}
```

在以上代码中，@ControllerAdvice 注解将注册 ErrorHandler 类，以应用于所有 Web 控制器。在 handleHttpStatusCodeException()方法中，如果出现 NotFoundException，则返回带有 HTTP 状态代码 404 的 ResponseEntity。

使用转换器，可以将使用@PathVariables 或@RequestParams 注解的 Web 控制器方法参数转换为值对象。而使用 HandlerMethodArgumentResolver，可以解析任何方法参数类型。

2.7 使用 Validation 实现 RESTful 服务的验证

在实际应用中，可能期望对 RESTful 服务有某种格式的请求，抑或是期望请求的元素具有某些数据类型、某些域约束。如果接收到不符合此限制的请求，怎么办？可以只返回通用消息吗？RESTful 服务的核心设计原则之一是：考虑消费者。因此，当请求中的某些内容无效时，应该返回正确的错误响应。清楚的响应消息会指出出现了什么问题、哪个字段有错误及可接受的值是什么、消费者可以采取什么措施来纠正错误等内容，正确地响应状态错误请求（响应中不要包含敏感信息）。验证错误的建议响应状态为->400 - Bad Request。

Spring Boot 为验证 RESTful 服务提供了良好的功能。下面就来快速了解其所提供的默认异常处理功能。

- 内容类型错误：如果将 Content-Type 用作 application/xml，但应用程序不支持此类型，则 Spring Boot 会默认返回 415 - Unsupported Media Type 的响应状态。
- 无效的 JSON 内容：如果将无效的 JSON 内容发送给需要正文的方法，则会返回 400 - Bad Request 的响应状态。
- 缺少元素的有效 JSON 结构：如果用户发送缺失或带有无效属性/元素的有效 JSON 结构，则应用程序将利用可用数据执行请求。

例如，以下请求的执行状态为-> 201 Created，这是将 POST 请求发送到 http://localhost:8080/students 的示例。

```
POST http://localhost:8080/students
```

请求内容（为空）：

```
{

}
```

再如，以下请求的执行状态为-> 201 Created，这是将 GET 请求发送到 http://localhost:8080/students 的示例。

```
GET http://localhost:8080/students
```

请求内容：

```
{
    "name": null,
    "passportNumber": "A12345678"
}
```

以下我们可能会注意到在上面的请求中具有无效的属性 name。以下是将 GET 请求发送到 http://localhost:8080/students 时的响应：

```
[ { "id": 1, "name": null, "passportNumber": null }, { "id": 2, "name": null, "passportNumber": "A12345678" }, { "id": 10001, "name": "Ranga", "passportNumber": "E1234567" }, { "id": 10002, "name": "Ravi", "passportNumber": "A1234568" } ]
```

可以看到，为 id 1 和 id 2 创建的两个资源对于不可用的值采用了 null，RESTful 服务忽略了无效的元素/属性。

为了自定义验证，我们将使用 Hibernate Validator，它是 Bean 验证 API 的一种实现方式。当使用 Spring Boot Starter Web 时，就可以获得 Hibernate Validator。

在添加验证前，需要添加一个依赖项，代码如下：

```
<dependency>
    <groupId>org.springframework.boot</groupId>
    <artifactId>spring-boot-starter-validation</artifactId>
</dependency>
```

下面向 Student 类的 Bean 添加一些验证。在这里，使用@Size 指定相应字段的最小长度，并在出现验证错误时返回一条指定消息。

```
@Entity
public class Student {
    @Id
    @GeneratedValue
    private Long id;

    @NotNull
    @Size(min=2, message="Name should have at least 2 characters")
    private String name;

    @NotNull
    @Size(min=7, message="Passport should have at least 7 characters")
    private String passportNumber;
    ...
}
```

Bean 验证 API 提供了许多诸如以下的注解。其中大多数注解的含义是不言自明的，因此这里不再详述。

- @DecimalMax
- @DecimalMin
- @Digits
- @Email
- @Future
- @FutureOrPresent
- @Max
- @Min
- @Negative
- @NegativeOrZero

- @NotBlank
- @NotEmpty
- @NotNull
- @Null
- @Past
- @PastOrPresent
- @Pattern
- @Positive
- @PositiveOrZero

为了在资源上启用验证，除了添加@RequestBody 以外，还需添加@Valid 注解。例如：

```
public ResponseEntity<Object> createStudent(@Valid @RequestBody Student student) {…}
```

当执行属性与约束不匹配的请求时，将默认返回 404 Bad 请求状态。但问题是没有返回任何指示出问题的细节。消费者知道发出了一个错误的请求，但是他们怎么知道哪个元素未通过验证，以及应该如何解决呢？

为了自定义验证响应，下面定义一个简单的错误响应 Bean。

```
public class ErrorDetails {
    private Date timestamp;
    private String message;
    private String details;

    public ErrorDetails(Date timestamp, String message, String details) {
        super();
        this.timestamp = timestamp;
        this.message = message;
        this.details = details;
    }
}
```

```
        :
    }
```

接下来，定义一个@ControllerAdvice 来处理验证错误。通过重写 ResponseEntityExceptionHandler{}中的 handleMethodArgumentNotValid(MethodArgumentNotValidException ex, HttpHeaders headers, HttpStatus status, WebRequest request)方法来做到这一点。

```
@ControllerAdvice
@RestController
public class CustomizedResponseEntityExceptionHandler extends ResponseEntityExceptionHandler {

    @Override
    protected ResponseEntity<Object> handleMethodArgumentNotValid(MethodArgumentNotValidException ex,HttpHeaders headers, HttpStatus status, WebRequest request) {
        ErrorDetails errorDetails = new ErrorDetails(new Date(), "Validation Failed",
        ex.getBindingResult().toString());
        return new ResponseEntity(errorDetails, HttpStatus.BAD_REQUEST);
    }
```

要使用 ErrorDetails 返回错误响应，需定义一个@ControllerAdvice，代码如下：

```
@ControllerAdvice
@RestController
public class CustomizedResponseEntityExceptionHandler extends ResponseEntityExceptionHandler {

    @ExceptionHandler(StudentNotFoundException)
    public final ResponseEntity<ErrorDetails> handleUserNotFoundException(StudentNotFoundException ex, WebRequest request) {
        ErrorDetails errorDetails = new ErrorDetails(new Date(), ex.getMessage(),
        request.getDescription(false));
        return new ResponseEntity<>(errorDetails, HttpStatus.NOT_FOUND);
    }
}
```

当执行属性与约束不匹配的请求时，将返回 404 Bad 请求状态。

请求内容：

```
{
    "name": "",
    "passportNumber": "A12345678"
}
```

此外，您还将获得一个响应体，指出出了什么问题。例如：

```
{
    "timestamp": 1512717715118,
    "message": "Validation Failed",
    "details": "org.springframework.validation.BeanPropertyBindingResult: 1 errors\nField error in object 'student' on field 'name': rejected value []; codes [Size.student.name,Size.name,Size.java.lang.String,Size]; arguments [org.springframework.context.support.DefaultMessageSourceResolvable: codes [student.name,name]; arguments []; default message [name],2147483647,2]; default message [Name should have at least 2 characters]"
}
```

2.8 启用 HTTPS

HTTPS 在 HTTP 的基础上通过传输加密和身份认证，从而保证了传输过程的安全性。SSL（Secure Socket Layer，安全套接层）是基于 HTTPS 下的一个协议加密层。

为了让网站支持 HTTPS，可以从腾讯云这样的网站申请 SSL 证书，得到 lietu.com.jks 和 keystorePass.txt 文件。然后修改 Tomcat 配置文件 server.xml，以增加对 HTTPS 的支持，代码如下：

```
<Connector
    protocol="org.apache.coyote.http11.Http11NioProtocol"
    port="443" maxThreads="200"
    scheme="https" secure="true" SSLEnabled="true"
    keystoreFile="E:/soft/apache-tomcat-9.0.13/conf/lietu.com.jks" keystorePass=
"941g4953970"
    clientAuth="false" sslProtocol="TLS"/>
```

2.9 本章小结

JSON 是一种与语言无关的数据格式。它源于 JavaScript，但许多编程语言包括生成和解析 JSON 格式数据的代码。curl 是一款支持 HTTP 的常用命令行工具，通过 curl 发送命令可以实现与 Web 服务交互。

此外，本章还介绍了 Spring Boot 整合 Swagger 2 生成接口文档，以及在 Web 服务器 Tomcat 中启用 HTTPS 的方法。

第 3 章

访问数据库

本章首先介绍 Spring Boot 持久性存储的 CRUD 操作,然后介绍在 Spring Boot 项目中使用 MyBatis 数据持久化框架等内容。

3.1 Spring Boot CRUD 操作

什么是 CRUD 操作?CRUD 操作代表创建(Create)、读取(Read)/检索(Retrieve)、更新(Update)和删除(Delete)操作。这些是持久性存储的 4 个基本功能。

CRUD 操作是实现动态网站的基础。在实际应用过程中,我们应该将 CRUD 与 HTTP 操作动词区分开。在 HTTP 操作下,如果要创建新记录,则应使用 HTTP 操作动词 POST;要更新记录,则应使用动词 PUT;如果要删除记录,则应使用动词 DELETE。而通过 CRUD 操作,用户和管理员有权在线检索、创建、编辑和删除记录。

在 Spring Boot 中,有许多执行 CRUD 操作的方式可供选择。其中,最为有效的方式是在 SQL 中创建一组存储过程来执行操作。CRUD 操作涉及在关系数据库应用程序中实现的所有主要功能。CRUD 所代表的每种操作都可以映射到 SQL 语句和 HTTP 方法。CRUD 操作对应表如表 3-1 所示。

表 3-1 CRUD 操作对应表

CRUD 操作	SQL 命令动词	HTTP 操作动词	RESTful Web 服务动词
Create	INSERT	PUT/POST	POST
Read	SELECT	GET	GET
Update	UPDATE	PUT/POST/PATCH	PUT
Delete	DELETE	DELETE	DELETE

Spring Boot 提供了一个名为 CrudRepository 的接口,其中包含用于 CRUD 操作的方法。CrudRepository 接口在包 org.springframework.data.repository 中被定义,它扩展了 Spring Data Repository 接口,在存储库上提供通用的 CRUD 操作。如果要在应用程序中使用 CrudRepository,则必须创建一个接口并扩展 CrudRepository。例如:

```
public interface StudentRepository extends CrudRepository<Student, Integer>
{
```

}

此外，JpaRepository扩展了PagingAndSortingRepository。它提供了实现分页的所有方法。

下面就设置一个Spring Boot应用程序并执行CRUD操作。

步骤1：打开Spring Initializr（http://start.spring.io）。

步骤2：选择Spring Boot 2.3.0.M1版本。

步骤3：提供群组名称，如com.javatpoint。

步骤4：提供工件ID，如spring-boot-crud-operation。

步骤5：添加依赖项Spring Web、Spring Data JPA和H2数据库。

步骤6：单击GENERATE（生成）按钮。

在包com.javatpoint.model中创建一个名为Books的模型类，代码如下：

```java
package com.javatpoint.model;
import javax.persistence.Column;
import javax.persistence.Entity;
import javax.persistence.Id;
import javax.persistence.Table;
//mark class as an Entity
@Entity
//defining class name as Table name
@Table
public class Books {
    //defining book id as primary key
    @Id
    @Column
    private int bookid;
    @Column
    private String bookname;
    @Column
    private String author;
    @Column
    private int price;

    public int getBookid() {
        return bookid;
    }
    public void setBookid(int bookid) {
        this.bookid = bookid;
    }
    public String getBookname() {
        return bookname;
    }
    public void setBookname(String bookname) {
        this.bookname = bookname;
    }
    public String getAuthor() {
        return author;
    }
    public void setAuthor(String author) {
        this.author = author;
```

```java
    }
    public int getPrice() {
        return price;
    }
    public void setPrice(int price) {
        this.price = price;
    }
}
```

在包 com.javatpoint.controller 中创建一个名为 BooksController 的控制器类代码如下:

```java
package com.javatpoint.controller;
import java.util.List;
import org.springframework.beans.factory.annotation.Autowired;
import org.springframework.web.bind.annotation.DeleteMapping;
import org.springframework.web.bind.annotation.GetMapping;
import org.springframework.web.bind.annotation.PathVariable;
import org.springframework.web.bind.annotation.PostMapping;
import org.springframework.web.bind.annotation.PutMapping;
import org.springframework.web.bind.annotation.RequestBody;
import org.springframework.web.bind.annotation.RestController;
import com.javatpoint.model.Books;
import com.javatpoint.service.BooksService;

//mark class as Controller
@RestController
public class BooksController {
    //autowire the BooksService class
    @Autowired
    BooksService booksService;

    //creating a GetMapping that retrieves all the books detail from the database
    @GetMapping("/book")
    private List<Books> getAllBooks() {
        return booksService.getAllBooks();
    }

    //creating a GetMapping that retrieves the detail of a specific book
    @GetMapping("/book/{bookid}")
    private Books getBooks(@PathVariable("bookid") int bookid) {
        return booksService.getBooksById(bookid);
    }

    //creating a DeleteMapping that deletes a specified book
    @DeleteMapping("/book/{bookid}")
    private void deleteBook(@PathVariable("bookid") int bookid) {
        booksService.delete(bookid);
    }

    //creating a PostMapping that posts the book detail in the database
    @PostMapping("/books")
```

```
    private int saveBook(@RequestBody Books books) {
        booksService.saveOrUpdate(books);
        return books.getBookid();
    }

    //creating a PutMapping that updates the book detail
    @PutMapping("/books")
    private Books update(@RequestBody Books books) {
        booksService.saveOrUpdate(books);
        return books;
    }
}
```

在包 com.javatpoint.service 中创建一个名为 BooksService 的服务类，代码如下：

```
package com.javatpoint.service;
import java.util.ArrayList;
import java.util.List;
import org.springframework.beans.factory.annotation.Autowired;
import org.springframework.stereotype.Service;
import com.javatpoint.model.Books;
import com.javatpoint.repository.BooksRepository;

//defining the business logic
@Service
public class BooksService {
    @Autowired
    BooksRepository booksRepository;

    //getting all books record by using the method findaAll() of CrudRepository
    public List<Books> getAllBooks() {
        List<Books> books = new ArrayList<Books>();
        booksRepository.findAll().forEach(books1 -> books.add(books1));
        return books;
    }

    //getting a specific record by using the method findById() of CrudRepository
    public Books getBooksById(int id) {
        return booksRepository.findById(id).get();
    }

    //saving a specific record by using the method save() of CrudRepository
    public void saveOrUpdate(Books books) {
        booksRepository.save(books);
    }

    //deleting a specific record by using the method deleteById() of CrudRepository
    public void delete(int id) {
        booksRepository.deleteById(id);
    }

    //updating a record
    public void update(Books books, int bookid) {
        booksRepository.save(books);
```

在包 com.javatpoint.repository 中创建一个名为 BooksRepository 的存储库接口，代码如下：

```
package com.javatpoint.repository;
import org.springframework.data.repository.CrudRepository;
//repository that extends CrudRepository
import com.javatpoint.model.Books;
public interface BooksRepository extends CrudRepository<Books, Integer>
{
}
```

在 application.properties 文件中配置数据源，代码如下：

```
spring.datasource.url=jdbc:h2:mem:books_data
spring.datasource.driverClassName=org.h2.Driver
spring.datasource.username=sa
spring.datasource.password=pwd
spring.jpa.database-platform=org.hibernate.dialect.H2Dialect
#enabling the H2 console
spring.h2.console.enabled=true
```

打开 SpringBootCrudOperationApplication.java 文件并将其作为 Java 应用程序运行，代码如下：

```
package com.javatpoint;
import org.springframework.boot.SpringApplication;
import org.springframework.boot.autoconfigure.SpringBootApplication;
@SpringBootApplication
public class SpringBootCrudOperationApplication {
    public static void main(String[] args) {
        SpringApplication.run(SpringBootCrudOperationApplication.class, args);
    }
}
```

使用 RESTful API 客户端 Postman 或者 Insomnia 在 Body 中插入以下数据。

```
{
    "bookid": "5433",
    "bookname": "Core and Advance Java",
    "author": "R. Nageswara Rao",
    "price": "800"
}
```

发送 GET 请求到 URL：http://localhost:8080/book，将会返回已插入数据库中的数据，如下所示。

```
[{"bookid":5433,"bookname":"Core and Advance Java","author":"R. Nageswara Rao","price":800}]
```

当然，也可以通过发送 DELETE 请求删除记录，还可以发送 PUT 请求来更新记录。

3.2 MyBatis 数据持久化框架

首先，使用 Spring Initializr 创建一个 Spring Boot 项目，然后添加 Web 依赖项，将 groupId

设置为com.lietu，将artifactId设置为boot-mybatis。

现在，应该有一个看起来像以下这样的带有POM的项目。

```xml
<?xml version="1.0" encoding="UTF-8"?>
<project xmlns="http://maven.apache.org/POM/4.0.0" xmlns:xsi="http://www.w3.org/2001/XMLSchema-instance"
    xsi:schemaLocation="http://maven.apache.org/POM/4.0.0 http://maven.apache.org/xsd/maven-4.0.0.xsd">
    <modelVersion>4.0.0</modelVersion>
    <parent>
        <groupId>org.springframework.boot</groupId>
        <artifactId>spring-boot-starter-parent</artifactId>
        <version>2.1.3.RELEASE</version>
        <relativePath/> <!-- lookup parent from repository -->
    </parent>
    <groupId>com.lietu</groupId>
    <artifactId>boot-mybatis</artifactId>
    <version>0.0.1-SNAPSHOT</version>
    <name>boot-mybatis</name>
    <description>Demo project for Spring Boot</description>

    <properties>
        <java.version>1.8</java.version>
    </properties>

    <dependencies>
        <dependency>
            <groupId>org.springframework.boot</groupId>
            <artifactId>spring-boot-starter-thymeleaf</artifactId>
        </dependency>
        <dependency>
            <groupId>org.springframework.boot</groupId>
            <artifactId>spring-boot-starter-web</artifactId>
        </dependency>
        <dependency>
            <groupId>org.mybatis.spring.boot</groupId>
            <artifactId>mybatis-spring-boot-starter</artifactId>
            <version>2.0.0</version>
        </dependency>

        <dependency>
            <groupId>org.springframework.boot</groupId>
            <artifactId>spring-boot-devtools</artifactId>
            <scope>runtime</scope>
        </dependency>
        <dependency>
            <groupId>mysql</groupId>
            <artifactId>mysql-connector-java</artifactId>
            <scope>runtime</scope>
        </dependency>
        <dependency>
            <groupId>org.projectlombok</groupId>
```

```xml
            <artifactId>lombok</artifactId>
            <optional>true</optional>
        </dependency>
        <dependency>
            <groupId>org.springframework.boot</groupId>
            <artifactId>spring-boot-starter-test</artifactId>
            <scope>test</scope>
        </dependency>
    </dependencies>
    <repositories>
        <repository>
            <id>aliyun</id>
            <url>http://maven.aliyun.com/nexus/content/groups/public/</url>
        </repository>
    </repositories>
</project>
```

创建数据库 springboot_mybatis，然后在该数据库中执行以下 SQL 语句。

```sql
DROP TABLE IF EXISTS 'user';

CREATE TABLE 'user' (
    'id' bigint(20) NOT NULL AUTO_INCREMENT COMMENT '主键',
    'username' varchar(255) DEFAULT NULL COMMENT '用户名',
    'password' varchar(255) DEFAULT NULL COMMENT '密码',
    'create_time' datetime DEFAULT NULL COMMENT '创建日期',
    PRIMARY KEY ('id')
) ENGINE=InnoDB AUTO_INCREMENT=5 DEFAULT CHARSET=utf8;
INSERT INTO'user'
(username, password, create_time)
VALUES('1', '1', '2020-1-1');
```

在 resources/下创建/mapper/UserMapperXML.xml，代码如下：

```xml
<?xml version="1.0" encoding="UTF-8" ?>
<!DOCTYPE mapper PUBLIC "-//mybatis.org//DTD Mapper 3.0//EN" "http://mybatis.org/dtd/mybatis-3-mapper.dtd" >
<mapper namespace="com.lietu.mapper.UserMapperXML">
    <select id="findAll" resultType="com.lietu.entity.User">
        select * from user
    </select>

    <select id="findById" resultType="com.lietu.entity.User">
        select * from user where id = #{id}
    </select>

    <insert id="save" parameterType="com.lietu.entity.User">
        insert into user(username,password,create_time) values(#{username},#{password},#{createTime})
    </insert>

    <update id="update" parameterType="com.lietu.entity.User">
        update user set username=#{username},password=#{password} where id=#{id}
    </update>
```

```xml
    <delete id="delete" parameterType="long">
        delete from user where id=#{id}
    </delete>
</mapper>
```

在包 com.lietu.controller 中添加一个名为 UserController 的控制器类，代码如下：

```java
package com.lietu.controller;

import com.lietu.entity.User;
import org.apache.ibatis.annotations.Param;
import org.springframework.beans.factory.annotation.Autowired;
import org.springframework.web.bind.annotation.GetMapping;
import com.lietu.mapper.UserMapperXML;
import org.springframework.web.bind.annotation.RestController;

@RestController
public class UserController {
    @Autowired
    private UserMapperXML usrMapper;
    @GetMapping("/getUsr")
    public User getUsr(@Param("id") Long id){
        User usr = usrMapper.findById(id);
        return usr;
    }
}
```

在启动类中注解扫描 mapper 接口，代码如下：

```java
package com.lietu;
import org.mybatis.spring.annotation.MapperScan;
import org.springframework.boot.SpringApplication;
import org.springframework.boot.autoconfigure.SpringBootApplication;

@SpringBootApplication
@MapperScan("com.lietu.mapper")
public class BootMybatisApplication {

    public static void main(String[] args) {
        SpringApplication.run(BootMybatisApplication.class, args);
    }

}
```

配置文件 application.properties 中添加 xml 地址，代码如下：

```
spring:
  datasource:
    url: jdbc:mysql://localhost:3306/springboot_mybatis?characterEncoding=utf-8&serverTimezone=UTC
    username: root
    password: password
    driver-class-name: com.mysql.cj.jdbc.Driver

#mybatis 配置
```

```yaml
mybatis:
  mapper-locations: classpath:mapper/**/*.xml
  type-aliases-package: com.lietu.entity
  configuration:
    # 使用 JDBC 的 getGeneratedKeys 可以获取数据库自增主键值
    use-generated-keys: true
    # 开启驼峰命名转换方式,例如 Table(create_time) -> Entity(createTime).这样,就不需要关心
如何进行字段匹配,mybatis 会自动识别大写字母和下画线
    map-underscore-to-camel-case: true


# 打印 sql
logging:
  level:
    com.lietu.mapper: DEBUG
```

在浏览器中输入 http://localhost:8080/getUsr?id=5 进行测试,返回结果如下:

```
{"id":5,"username":"1","password":"1","createTime":"2020-01-01T00:00:00.000+0000"}
```

3.3 使用 HikariCP 连接池

HikariCP 是一款快速、简单、可靠的 JDBC 连接池。如果使用的是 Spring Boot 2.0 或其以后的版本,则不需要单独在 pom.xml 文件中引入 HikariCP 依赖。可以像以下这样使用 HikariConfig 类:

```java
HikariConfig config = new HikariConfig();
config.setJdbcUrl("jdbc:mysql://localhost:3306/simpsons");
config.setUsername("bart");
config.setPassword("51mp50n");
config.addDataSourceProperty("cachePrepStmts", "true");
config.addDataSourceProperty("prepStmtCacheSize", "250");
config.addDataSourceProperty("prepStmtCacheSqlLimit", "2048");
HikariDataSource ds = new HikariDataSource(config);
```

或者,像以下这样根据属性文件创建 HikariConfig 对象。

```java
//同时从文件系统和类路径检查.properties 文件
HikariConfig config = new HikariConfig("/some/path/hikari.properties");
HikariDataSource ds = new HikariDataSource(config);
```

属性文件的内容如下:

```
dataSourceClassName=org.postgresql.ds.PGSimpleDataSource
dataSource.user=test
dataSource.password=test
dataSource.databaseName=mydb
dataSource.portNumber=5432
dataSource.serverName=localhost
```

在 Spring Boot 1.3.0 中,可以配置 application.yml 文件。其内容如下:

```yaml
spring:
  datasource:
    type: com.zaxxer.hikari.HikariDataSource
```

```
    url: jdbc:h2:mem:TEST
    driver-class-name: org.h2.Driver
    username: username
    password: password
    hikari:
      idle-timeout: 10000
```

3.4 缓存

缓存有助于通过减少数据库或任何昂贵资源之间的往返访问/调用次数来提高应用程序的性能。在日常应用中，将随时要面对必须执行大量数据库查询等的场景，即使假设数据库中的数据很少会发生变化的情况下，每次都调用数据库也不是一个好主意。面对这种场景，可以只缓存第一次调用数据库时的结果，并为其他调用再次返回相同的数据。

为了在 Spring Boot 应用程序中使用缓存，可以先将@EnableCaching 注解写入主类，然后将@Cacheable 注解添加到要缓存结果的方法中。

例如，要缓存文章。创建文章的实体类 Article，其代码如下：

```
public class Article implements Serializable {
    private long articleId;
    private String title;
    private String category;

    public Article(long i, String t, String c) {
        articleId = i;
        title = t;
        category = c;
    }
}
```

在以下的代码中，@EnableCaching 支持注解驱动的缓存管理功能，它负责注册所需的 Spring Boot 组件以启用注解驱动的缓存管理。@EnableCaching 使用@Configuration 或@SpringBootApplication 注解进行注解。CacheExample 主类的实现代码如下：

```
@RestController
@SpringBootApplication
@EnableCaching
public class CacheExample extends SpringBootServletInitializer {
    @RequestMapping("/get-article-info")
    @Cacheable(value="cacheArticleInfo")
    public ResponseEntity<List<Article>> articleInformation() {
        System.out.println("get articleInformation");
        List<Article> articleDetails = Arrays.asList(
            /*
            *在这里,可以添加数据库逻辑/流程以获取文章的详细信息
            */
            new Article(100,"title1","content1"),
            new Article(101,"title2","content2")
```

```
            );
        return ResponseEntity.ok(articleDetails);
    }

    public static void main(String[] args) {
        SpringApplication.run(Example.class, args);
    }
}
```

通过网址（http://localhost:8080/get-article-info）测试缓存效果。

RedisCacheManager 是一个由 Spring Boot 提供的 Redis 所支持的 CacheManager。如果在我们的应用程序类路径中提供了 Redis 并且所需的配置可用，则 Spring Boot 会自动配置 RedisCacheManager 实例。

在 Spring Boot 中还提供了 spring-boot-starter-data-redis 来解决 Redis 依赖关系问题，并为 Lettuce 和 Jedis 客户端库提供基本的自动配置。默认情况下，Spring Boot 2.0 使用 Lettuce。要获得池化连接工厂，需要提供 commons-pool2 依赖项。如果使用 Lettuce，需要如下 Gradle 依赖项。

```
dependencies {
    compile group: 'org.springframework.boot', name: 'spring-boot-starter-data-redis', version: '2.1.3.RELEASE'
    compile group: 'org.apache.commons', name: 'commons-pool2', version: '2.6.1'
}
```

要配置 Lettuce 池，需要将 spring.redis.* 前缀与 Lettuce 池连接属性一起使用。Lettuce 池样本配置如下：

```
spring.redis.host=localhost
spring.redis.port=6379
spring.redis.password=
spring.redis.lettuce.pool.max-active=7
spring.redis.lettuce.pool.max-idle=7
spring.redis.lettuce.pool.min-idle=2
spring.redis.lettuce.pool.max-wait=-1ms
spring.redis.lettuce.shutdown-timeout=200ms
```

在以上配置代码中，可以覆盖默认的 Redis 主机、端口和密码配置。如果想要无限期地阻止，可以设置 max-wait 为一个负值。

此外，还可以使用 spring.cache.* 属性控制 Spring 缓存配置。例如：

```
spring.cache.redis.cache-null-values=false
spring.cache.redis.time-to-live=600000
spring.cache.redis.use-key-prefix=true
spring.cache.type=redis
spring.cache.cache-names=articleCache,allArticlesCache
```

在以上配置代码中，缓存 articleCache 和 allArticlesCache 将存活 10min（即 600 000ms）。

我们可以创建自己的 RedisCacheManager 来完全控制 Redis 配置。我们需要创建 Lettuce ConnectionFactory bean、RedisCacheConfiguration bean 和 RedisCacheManager，代码如下：

```
@Configuration
```

```java
@EnableCaching
@PropertySource("classpath:application.properties")
public class RedisConfig {
   @Autowired
   private Environment env;

   @Bean
   public LettuceConnectionFactory redisConnectionFactory() {
   RedisStandaloneConfiguration redisConf = new RedisStandaloneConfiguration();
   redisConf.setHostName(env.getProperty("spring.redis.host"));
   redisConf.setPort(Integer.parseInt(env.getProperty("spring.redis.port")));
   redisConf.setPassword(RedisPassword.of(env.getProperty("spring.redis.password")));
   return new LettuceConnectionFactory(redisConf);
   }
   @Bean
   public RedisCacheConfiguration cacheConfiguration() {
   RedisCacheConfiguration cacheConfig = RedisCacheConfiguration.defaultCacheConfig()
     .entryTtl(Duration.ofSeconds(600))
     .disableCachingNullValues();
   return cacheConfig;
   }
   @Bean
   public RedisCacheManager cacheManager() {
   RedisCacheManager rcm = RedisCacheManager.builder(redisConnectionFactory())
     .cacheDefaults(cacheConfiguration())
     .transactionAware()
     .build();
   return rcm;
   }
}
```

这里的 RedisCacheConfiguration 是不可变类，可帮助自定义 Redis 缓存行为，如缓存到期时间及禁用缓存空值等。它还有助于自定义序列化策略。

如果要禁用缓存，则无须删除所有注释。只需在 application.properties 文件中添加以下语句，Spring Boot 会自动为你完成相应操作。

```
spring.cache.type=none
```

3.5 MongoDB 数据库

MongoDB 是一款文档数据库，具有所需的可伸缩性和灵活性，可用于查询和索引。应用 MongoDB 以前，需要在操作系统中安装 MongoDB。首先从 MongoDB 下载中心下载 MongoDB 服务器社区版，在 Windows 操作系统下安装后，通过从 cmd 窗口执行以下命令来运行 MongoDB 服务器。

```
E:\Program Files\mongodb\Server\4.2\bin>mongod.exe
```

此时，MongoDB 服务器在默认端口上运行：27017。

接着，执行以下命令来运行 MongoDB 客户端。

```
E:\Program Files\mongodb\Server\4.2\bin>mongo.exe
```

在 MongoDB 客户端上执行以下命令创建新数据库。

```
> use EmployeeDB
switched to db EmployeeDB
```

检查数据库:

```
> db
EmployeeDB
```

显示所有数据库:

```
> show dbs
admin   0.000GB
config  0.000GB
local   0.000GB
```

列出当前数据库的集合:

```
> show collections;
```

创建一个集合:

```
db.createCollection("collectionName");
```

将文档插入集合中:

```
//插入单个文档
//
db.<collectionName>.insert({field1: "value", field2: "value"})
//
//插入多个文档
//
db.<collectionName>.insert([{field1: "value1"}, {field1: "value2"}])
db.<collectionName>.insertMany([{field1: "value1"}, {field1: "value2"}])
```

保存或更新文档:

```
//匹配文件将被更新;如果找不到与 ID 匹配的文档,则创建一个新文档
db.<collectionName>.save({"_id": new ObjectId("jhgsdjhgdsf"), field1: "value", field2: "value"});
```

显示集合记录:

```
//
//检索所有记录
//
db.<collectionName>.find();
//
//检索有限数量的记录;以下命令将打印 10 条结果
//
db.<collectionName>.find().limit(10);
//
//按 ID 检索记录
//
db.<collectionName>.find({"_id": ObjectId("someid")});
//
//通过传递具有以下内容的对象来检索指定集合属性的值
//根据属性值是否需要包含在输出中来给属性名称分配 1 或 0
db.<collectionName>.find({"_id": ObjectId("someid")}, {field1: 1, field2: 1});
```

```
db.<collectionName>.find({"_id": ObjectId("someid")}, {field1: 0}); //排除field1
//
//统计集合中的文档数量
//
db.<collectionName>.count();
```

索引上的 MongoDB 查询不是原子性的，并且可能会错过在查询运行时正在更新的文档，并在文档更新前和更新后匹配查询。

管理命令：

```
//
//获取集合统计信息
//
db.<collectionName>.stats()
db.printCollectionStats()
//
//读取、写入操作的延迟统计信息,包括读取、写入所需的平均时间,以及相关的操作数量
//
db.<collectionName>.latencyStats()
//
//获取数据和索引的集合大小
//
db.<collectionName>.dataSize()           //集合的大小
db.<collectionName>.storageSize()        //集合中存储的文档总大小
db.<collectionName>.totalSize()          //集合数据和索引的总大小（以"字节"为单位）
db.<collectionName>.totalIndexSize()     //集合中所有索引的总大小
```

如果要在 Linux 操作系统下安装 MongoDB，则首先执行以下命令来下载安装文件。

```
curl -O https://fastdl.mongodb.org/linux/mongodb-linux-x86_64-3.4.4.tgz
```

然后执行以下命令来解压缩该文件。

```
tar -zxvf mongodb-linux-x86_64-3.4.4.tgz
```

在首次启动 MongoDB 前，需要创建 mongod 进程写入数据的目录。默认情况下，mongod 会将数据写入/data/db 目录。使用以下命令创建该目录：

```
mkdir -p /data/db
```

运行可执行文件 mongod：

```
<path to binary>/mongod
```

停止运行：

```
# ./mongod --shutdown
```

创建 MongoDB 服务：

```
# mkdir ../log/
# ./mongod --fork --logpath ../log/mongod.log
about to fork child process, waiting until server is ready for connections.
forked process: 30183
child process started successfully, parent exiting
```

存储数据的默认路径为/data/db，可以执行以下命令来显示该目录下文件的详细信息。

```
# ls /data/db
```

在配置文件 mongod.conf 中执行以下命令定义绑定的 IP 地址，让 mongod 能够接收外部连接。

```
net:
   bindIp: 0.0.0.0
   port: 27017
```

使用配置文件：

```
# ./mongod -f /etc/mongod.conf
```

可以通过 mongoimport 命令把数据导入到 local 数据库。例如：

```
# ./mongoimport -d local -c users --file ./sample_2.json --type json
```

相应地，也可以用 mongoexport 命令把数据导出。

①查看有关该命令的帮助信息：

```
#./mongoexport --help
```

②导出数据：

```
# ./mongoexport -d local -c ship -o ship.json
```

MongoDB 启动警告：soft rlimits too low. rlimits set to 15082 processes, 65535 files. Number of processes should be at least 32767.5 : 0.5 times number of files.

查看最大进程数量：

```
#sysctl kernel.pid_max
```

在命令行设置：

```
# sysctl kernel.pid_max=4194303
```

在配置文件/etc/sysctl.conf 中设置：

```
kernel.pid_max = 4194303
```

可以通过 JDBC 连接到 MongoDB 数据库，代码如下：

```java
MongoClient mongoClient = new MongoClient("localhost", 27017); //或添加连接的 IP 地址
MongoDatabase db = mongoClient.getDatabase("local");
System.out.println("Connect to database successfully");
// boolean auth = db.authenticate(myUserName, myPassword);
// System.out.println("Authentication: "+auth);
MongoCollection<Document> coll = db.getCollection("sample_2");
FindIterable<Document> cursor = coll.find();
int i = 0;
for (Iterator<Document> iterator = cursor.iterator(); iterator.hasNext(); i++) {
    System.out.println(iterator.next());
    if (i == 10)
        break;
}
mongoClient.close();
```

Document.append("key", Value)添加值到 Document 的字段中，代码如下：

```java
Document personObj = new Document("name", "john")
        .append("age", 35)
        .append("info", new BasicDBObject("email", "john@mail.com")
                .append("phone", "876-134-667"));
```

```
coll.insertOne(personObj);
```

插入嵌套对象：

```
List<DBObject> kids = Arrays.<DBObject> asList(new BasicDBObject(
    "firstName", "john").append("lastName", "doe"),
    new BasicDBObject("firstName", "anna").append("lastName", "smith"),
    new BasicDBObject("firstName", "peter") .append("lastName", "jones"));
Document personObj = new Document("name", "john")
    .append("age", 35)
    .append("kids", kids)
    .append("info", new BasicDBObject("email", "john@mail.com")
        .append("phone", "876-134-667"));

coll.insertOne(personObj);
```

用 limit 方法查找前 10 个文档，代码如下：

```
FindIterable<Document> cursor = coll.find().limit(10);
for (Iterator<Document> iterator = cursor.iterator(); iterator.hasNext(); ) {
    System.out.println(iterator.next());
}
```

排序要传入一个 Bson 对象，代码如下：

```
Bson sort = new BasicDBObject("name_eng", 1); //升序
FindIterable<Document> cursor = coll.find().sort(sort).limit(10);
```

选择列代码如下：

```
Bson fields = new BasicDBObject("full_name", true);
FindIterable<Document> cursor = coll.find().projection(fields).limit(10);
```

删除表代码如下：

```
String cName = "ship";
MongoCollection<Document> coll = db.getCollection(cName);

coll.drop(); //删除表
```

注意：

① 非必要时，避免使用 where 子句。因为它们在速度方面比常规查询慢很多，每个文档都要从 Bson 对象转换为 JS 对象后运行。

② 可以清空 MongoDB 中 n 天之前的数据。为了避免自己编写脚本去清除这些麻烦的数据，可以使用 TTL 索引让这些数据保存 n 天，也就是 n 天之后自动删除。TTL（Time To Live）索引是特定的单字段索引，在 MongoDB 中利用该索引方式可以实现在特定时间或特定时间段后自动从集合中删除文档。数据到期对于某些类型的信息（如机器生成的事件数据、日志和会话信息）处理很有用，因为这样的信息只需要在数据库中持续有限的时间。要创建 TTL 索引，可以在其值为日期或包含日期值的数组字段上使用 db.collection.createIndex()方法，再加上 expireAfterSeconds 选项。例如，要在 eventlog 集合的 lastModifiedDate 字段上创建 TTL 索引，可以在 mongo Shell 中执行以下操作。

```
db.eventlog.createIndex( { "lastModifiedDate": 1 }, { expireAfterSeconds: 3600 } )
```

在下面的 pom.xml 文件中定义构建此应用程序所需的依赖项。

```xml
<?xml version="1.0" encoding="UTF-8"?>
<project
    xmlns="http://maven.apache.org/POM/4.0.0"
    xmlns:xsi="http://www.w3.org/2001/XMLSchema-instance"
    xsi:schemaLocation="http://maven.apache.org/POM/4.0.0 https://maven.apache.org/xsd/maven-4.0.0.xsd">
    <modelVersion>4.0.0</modelVersion>
    <parent>
        <groupId>org.springframework.boot</groupId>
        <artifactId>spring-boot-starter-parent</artifactId>
        <version>2.1.0.RELEASE</version>
        <relativePath />
        <!-- lookup parent from repository -->
    </parent>
    <groupId>net.springboot.javaguides</groupId>
    <artifactId>springboot-mongodb-crud-example</artifactId>
    <version>0.0.1-SNAPSHOT</version>
    <name>springboot-thymeleaf-web-app</name>
    <description>Demo project for Spring Boot + Mongo DB CRUD Example</description>
    <properties>
        <java.version>1.8</java.version>
    </properties>
    <dependencies>
        <dependency>
            <groupId>org.springframework.boot</groupId>
            <artifactId>spring-boot-starter-data-mongodb</artifactId>
        </dependency>
        <dependency>
            <groupId>org.springframework.boot</groupId>
            <artifactId>spring-boot-starter-web</artifactId>
        </dependency>
        <dependency>
            <groupId>org.springframework.boot</groupId>
            <artifactId>spring-boot-starter-test</artifactId>
            <scope>test</scope>
            <exclusions>
                <exclusion>
                    <groupId>org.junit.vintage</groupId>
                    <artifactId>junit-vintage-engine</artifactId>
                </exclusion>
            </exclusions>
        </dependency>
    </dependencies>
    <build>
        <plugins>
            <plugin>
                <groupId>org.springframework.boot</groupId>
                <artifactId>spring-boot-maven-plugin</artifactId>
            </plugin>
```

```
        </plugins>
    </build>
</project>
```

Spring Boot 会根据以上在 pom.xml 文件中添加的依赖项，尝试自动配置大多数内容。例如，由于添加了 spring-boot-starter-data-mongodb 依赖关系，因此 Spring Boot 会尝试通过从 application.properties 文件中读取数据库配置来建立与 MongoDB 的连接。

打开 application.properties 文件，并添加以下 MongoDB 属性。

```
# MONGODB (MongoProperties)
spring.data.mongodb.uri=mongodb://localhost:27017/EmployeeDB
```

根据配置，MongoDB 在默认端口 27017 上实现本地运行。注意，需要使用以下命令创建 EmployeeDB 数据库。

```
use EmployeeDB
```

如果 MongoDB 中不存在 EmployeeDB 数据库，则它将创建一个新数据库。

下面创建一个集合 database_sequences。该集合用于存储其他集合的自动递增序列，可以使用 mongo Shell 或 MongoDB Compass 管理工具创建。

首先，创建一个对应的模型类，代码如下：

```
package net.guides.springboot.crud.model;
import org.springframework.data.annotation.Id;
import org.springframework.data.mongodb.core.mapping.Document;

@Document(collection = "database_sequences")
public class DatabaseSequence {
    @Id
    private String id;
    private long seq;
    public DatabaseSequence() { }
    public String getId() {
        return id;
    }
    public void setId(String id) {
        this.id = id;
    }
    public long getSeq() {
        return seq;
    }
    public void setSeq(long seq) {
        this.seq = seq;
    }
}
```

然后，创建 Employee 模型，该模型将映射到 MongoDB 数据库中的文档。在 net.guides.springboot.crud 包内创建一个新的 model 包，并在 model 包内添加一个具有以下代码的文件 Employee.java。

```
package net.guides.springboot.crud.model;
import javax.validation.constraints.NotBlank;
import javax.validation.constraints.Size;
import org.springframework.data.annotation.Id;
```

```java
import org.springframework.data.annotation.Transient;
import org.springframework.data.mongodb.core.index.Indexed;
import org.springframework.data.mongodb.core.mapping.Document;

@Document(collection = "Employee")
public class Employee {
   @Transient
   public static final String SEQUENCE_NAME = "users_sequence";
   @Id
   private long id;
   @NotBlank
   @Size(max = 100)
   @Indexed(unique = true)
   private String firstName;
   private String lastName;
   @NotBlank
   @Size(max = 100)
   @Indexed(unique = true)
   private String emailId;
   public Employee() {
   }
   public Employee(String firstName, String lastName, String emailId) {
      this.firstName = firstName;
      this.lastName = lastName;
      this.emailId = emailId;
   }
   public long getId() {
      return id;
   }
   public void setId(long id) {
      this.id = id;
   }

   public String getFirstName() {
      return firstName;
   }
   public void setFirstName(String firstName) {
      this.firstName = firstName;
   }
   public String getLastName() {
      return lastName;
   }
   public void setLastName(String lastName) {
      this.lastName = lastName;
   }
   public String getEmailId() {
      return emailId;
   }
   public void setEmailId(String emailId) {
      this.emailId = emailId;
   }
```

```
    @Override
    public String toString() {
        return "Employee [id=" + id + ", firstName=" + firstName + ", lastName=" +
lastName + ", emailId=" + emailId + "]";
    }
}
```

在以上代码中,有一个带@Indexed 注解的 firstName 和一个带有@Indexed 注解的 emailId,并将它们标记为唯一。 这将在 firstName 和 emailId 字段上创建唯一索引。

接下来,需要创建 EmployeeRepository 来访问数据库中的数据,代码如下:

```
package net.guides.springboot.crud.repository;
import org.springframework.data.mongodb.repository.MongoRepository;
import org.springframework.stereotype.Repository;
import net.guides.springboot.crud.model.Employee;

@Repository
public interface EmployeeRepository extends MongoRepository<Employee, Long>{
}
```

创建 SequenceGeneratorService 来实现自动生成序列,代码如下:

```
package net.guides.springboot.crud.service;
import static org.springframework.data.mongodb.core.FindAndModifyOptions.options;
import static org.springframework.data.mongodb.core.query.Criteria.where;
import static org.springframework.data.mongodb.core.query.Query.query;
import java.util.Objects;
import org.springframework.beans.factory.annotation.Autowired;
import org.springframework.data.mongodb.core.MongoOperations;
import org.springframework.data.mongodb.core.query.Update;
import org.springframework.stereotype.Service;
import net.guides.springboot.crud.model.DatabaseSequence;

@Service
public class SequenceGeneratorService {
    private MongoOperations mongoOperations;
    @Autowired
    public SequenceGeneratorService(MongoOperations mongoOperations) {
        this.mongoOperations = mongoOperations;
    }
    public long generateSequence(String seqName) {
        DatabaseSequence counter = mongoOperations.findAndModify(query(where("_id").
is(seqName)),
            new Update().inc("seq",1), options().returnNew(true).upsert(true),
            DatabaseSequence.class);
        return !Objects.isNull(counter) ? counter.getSeq() : 1;
    }
}
```

创建将向客户端公开的 API。在这里需注意的是,注解@RestController、@RequestMapping、@GetMapping、@PostMapping 和@DeleteMapping 将各种 URI 映射到控制器方法的用法。注意,在创建新记录时正在使用 generateSequence()方法,代码如下:

```
package net.guides.springboot.crud.controller;
```

```java
import java.util.HashMap;
import java.util.List;
import java.util.Map;
import javax.validation.Valid;
import org.springframework.beans.factory.annotation.Autowired;
import org.springframework.http.ResponseEntity;
import org.springframework.web.bind.annotation.CrossOrigin;
import org.springframework.web.bind.annotation.DeleteMapping;
import org.springframework.web.bind.annotation.GetMapping;
import org.springframework.web.bind.annotation.PathVariable;
import org.springframework.web.bind.annotation.PostMapping;
import org.springframework.web.bind.annotation.PutMapping;
import org.springframework.web.bind.annotation.RequestBody;
import org.springframework.web.bind.annotation.RequestMapping;
import org.springframework.web.bind.annotation.RestController;
import net.guides.springboot.crud.exception.ResourceNotFoundException;
import net.guides.springboot.crud.model.Employee;
import net.guides.springboot.crud.repository.EmployeeRepository;
import net.guides.springboot.crud.service.SequenceGeneratorService;

//@CrossOrigin(origins = "http://localhost:4200")
@RestController
@RequestMapping("/api/v1")
public class EmployeeController {
    @Autowired
    private EmployeeRepository employeeRepository;

    @Autowired
    private SequenceGeneratorService sequenceGeneratorService;

    @GetMapping("/employees")
    public List < Employee > getAllEmployees() {
        return employeeRepository.findAll();
    }

    @GetMapping("/employees/{id}")
    public ResponseEntity < Employee > getEmployeeById(@PathVariable(value = "id") Long employeeId)
    throws ResourceNotFoundException {
        Employee employee = employeeRepository.findById(employeeId)
            .orElseThrow(() -> new ResourceNotFoundException("Employee not found for this id :: " + employeeId));
        return ResponseEntity.ok().body(employee);
    }

    @PostMapping("/employees")
    public Employee createEmployee(@Valid @RequestBody Employee employee) {
        employee.setId(sequenceGeneratorService.generateSequence(Employee.SEQUENCE_NAME));
        return employeeRepository.save(employee);
    }
```

```
    @PutMapping("/employees/{id}")
    public ResponseEntity < Employee > updateEmployee(@PathVariable(value = "id") Long employeeId,
        @Valid @RequestBody Employee employeeDetails) throws ResourceNotFoundException {
        Employee employee = employeeRepository.findById(employeeId)
            .orElseThrow(() - > new ResourceNotFoundException("Employee not found for this id :: " + employeeId));
        employee.setEmailId(employeeDetails.getEmailId());
        employee.setLastName(employeeDetails.getLastName());
        employee.setFirstName(employeeDetails.getFirstName());
        final Employee updatedEmployee = employeeRepository.save(employee);
        return ResponseEntity.ok(updatedEmployee);
    }

    @DeleteMapping("/employees/{id}")
    public Map < String, Boolean > deleteEmployee(@PathVariable(value = "id") Long employeeId) throws ResourceNotFoundException {
        Employee employee = employeeRepository.findById(employeeId)
            .orElseThrow(() - > new ResourceNotFoundException("Employee not found for this id :: " + employeeId));
        employeeRepository.delete(employee);
        Map < String, Boolean > response = new HashMap < > ();
        response.put("deleted", Boolean.TRUE);
        return response;
    }
}
```

这个 Spring Boot 应用程序具有一个名为 Application.java 的入口点 Java 类，在其代码中应用了 public static void main(String[] args)方法，可以运行该方法来启动该应用程序，代码如下：

```
import org.springframework.boot.SpringApplication;
import org.springframework.boot.autoconfigure.SpringBootApplication;

@SpringBootApplication
public class Application {
    public static void main(String[] args) {
        SpringApplication.run(Application.class, args);
    }
}
```

3.6 本章小结

2001 年，克林顿·贝恩（Clinton Begin）启动了一个名为 iBATIS 的项目。最初的重点是密码软件解决方案的开发。iBATIS 发布的第一个产品是 Secrets，这是一个个人数据加密和签名工具。在发布适用于 Java 的 Secrets 之后不久，iBATIS 项目出现了转机，并开始专注于 Web 和其他与 Internet 相关的技术。MyBatis 是 iBATIS 3.0 的一个分支，由包括 iBATIS 原始创建者的团队维护。

MongoDB 的存储引擎负责管理数据如何存储在硬盘和内存上。从 MongoDB 3.2 开始，

WiredTiger 存储引擎是默认的存储引擎。

本章首先介绍了使用 CrudRepository 接口实现持久性存储，通过示例介绍了 MyBatis 数据持久化框架，HikariCP 数据库连接池的使用，然后介绍了在 SpringBoot 应用程序中使用缓存，最后介绍了通过 Spring Data MongoDB 使用 MongoDB 数据库。

第 4 章 权限管理

本章首先介绍 Spring Security 实现的权限控制，然后介绍 Apache Shiro 实现的权限控制及集成 JWT 身份验证，最后介绍 OAuth 2 授权框架。

4.1 Security 实现权限控制

在 Spring Boot 中，可以使用 Spring Boot Starter Security 来启动 Web 应用程序的安全性。Spring Security 默认情况下会保护所有 HTTP 端点，而用户必须以默认的 HTTP 形式登录。

项目的 pom.xml 文件代码如下：

```xml
<?xml version="1.0" encoding="UTF-8"?>
<project xmlns="http://maven.apache.org/POM/4.0.0"
    xmlns:xsi="http://www.w3.org/2001/XMLSchema-instance"
    xsi:schemaLocation="http://maven.apache.org/POM/4.0.0 http://maven.apache.org/xsd/maven-4.0.0.xsd">
    <modelVersion>4.0.0</modelVersion>
    <groupId>com.zetcode</groupId>
    <artifactId>springbootloginpage</artifactId>
    <version>1.0-SNAPSHOT</version>
    <packaging>jar</packaging>

    <properties>
        <project.build.sourceEncoding>UTF-8</project.build.sourceEncoding>
        <maven.compiler.source>11</maven.compiler.source>
        <maven.compiler.target>11</maven.compiler.target>
    </properties>

    <parent>
        <groupId>org.springframework.boot</groupId>
        <artifactId>spring-boot-starter-parent</artifactId>
        <version>2.1.5.RELEASE</version>
    </parent>

    <dependencies>
        <dependency>
            <groupId>org.springframework.boot</groupId>
            <artifactId>spring-boot-starter-web</artifactId>
        </dependency>
```

```xml
        <dependency>
            <groupId>org.springframework.boot</groupId>
            <artifactId>spring-boot-starter-security</artifactId>
        </dependency>

    </dependencies>

    <build>
        <plugins>
            <plugin>
                <groupId>org.springframework.boot</groupId>
                <artifactId>spring-boot-maven-plugin</artifactId>
            </plugin>
        </plugins>
    </build>
</project>
```

注意：在上述代码中，增加了 Web 和安全性的 starter 依赖项。

resources/application.properties 文件代码如下：

```
spring.main.banner-mode=off
logging.pattern.console=%d{dd-MM-yyyy HH:mm:ss} %magenta([%thread]) %highlight(%-5level) %logger.%M - %msg%n
```

在以上 application.properties 文件中，关闭 Spring Boot banner 并配置控制台日志记录模式。

com/lietu/controller/MyController.java 文件代码如下：

```java
package com.lietu.controller;

import org.springframework.web.bind.annotation.GetMapping;
import org.springframework.web.bind.annotation.RestController;

@RestController
public class MyController {

    @GetMapping("/")
    public String home() {
        return "This is home page";
    }
}
```

该应用程序有一个简单的主页。运行该应用程序，并导航到 localhost:8080，界面会被重定向到 http://localhost:8080/login 页面。

运行返回信息如下：

```
...
17-06-2019 17:48:45 [main] INFO org.springframework.boot.autoconfigure.security.servlet.UserDetailsServiceAutoConfiguration.getOrDeducePassword -
Using generated security password: df7ce50b-abae-43a1-abe1-0e17fd81a454
...
```

在控制台中，可以看到名为 user 的默认用户生成的密码。这些凭据将提供给身份验证表单。Spring Boot 使用 Bootstrap 框架定义登录 UI。使用以下语句就可以实现拥有新的用户名和密码。同时，使用这些设置将关闭自动生成账户功能。

```
spring.security.user.name = admin
spring.security.user.password = s$cret
```

4.2　Shiro 实现权限控制

Apache Shiro 是一款功能强大且易于使用的 Java 安全框架，它可以执行身份验证、授权、加密和会话管理等操作。

Spring Boot 集成 Shiro 项目的 pom.xml 文件代码如下：

```xml
<?xml version="1.0" encoding="UTF-8"?>
<project xmlns="http://maven.apache.org/POM/4.0.0" xmlns:xsi="http://www.w3.org/2001/XMLSchema-instance"
    xsi:schemaLocation="http://maven.apache.org/POM/4.0.0 http://maven.apache.org/xsd/maven-4.0.0.xsd">
    <modelVersion>4.0.0</modelVersion>
    <parent>
        <groupId>org.springframework.boot</groupId>
        <artifactId>spring-boot-starter-parent</artifactId>
        <version>2.2.5.RELEASE</version>
        <relativePath/> <!-- lookup parent from repository -->
    </parent>
    <groupId>com.lietu</groupId>
    <artifactId>shiro</artifactId>
    <version>0.0.1-SNAPSHOT</version>
    <name>shiro</name>
    <description>Demo project for Spring Boot</description>

    <properties>
        <java.version>1.8</java.version>
    </properties>

    <dependencies>
        <dependency>
            <groupId>org.springframework.boot</groupId>
            <artifactId>spring-boot-starter-thymeleaf</artifactId>
        </dependency>
        <!--shiro 1.4.0 thymeleaf-extras-shiro 2.0.0 组合-->
        <dependency>
            <groupId>org.apache.shiro</groupId>
            <artifactId>shiro-core</artifactId>
            <version>1.4.0</version>
        </dependency>
        <dependency>
            <groupId>org.apache.shiro</groupId>
            <artifactId>shiro-spring</artifactId>
            <version>1.4.0</version>
        </dependency>
```

```xml
        <!--shiro for thymeleaf 生效需要加入 Spring Boot 2.x 版本,请使用 2.0.0 版本或使用
1.2.1版本-->
        <dependency>
            <groupId>com.github.theborakompanioni</groupId>
            <artifactId>thymeleaf-extras-shiro</artifactId>
            <version>2.0.0</version>
        </dependency>
        <dependency>
            <groupId>org.springframework.boot</groupId>
            <artifactId>spring-boot-starter-web</artifactId>
        </dependency>
        <dependency>
            <groupId>org.springframework.boot</groupId>
            <artifactId>spring-boot-starter-test</artifactId>
            <scope>test</scope>
        </dependency>
    </dependencies>

    <build>
        <plugins>
            <plugin>
                <groupId>org.springframework.boot</groupId>
                <artifactId>spring-boot-maven-plugin</artifactId>
            </plugin>
        </plugins>
    </build>

    <repositories>
        <repository>
            <id>aliyun</id>
            <url>http://maven.aliyun.com/nexus/content/groups/public/</url>
        </repository>
    </repositories>
</project>
```

在配置文件 application.yml 中修改端口号,代码如下:

```yaml
server:
  port: 8086
```

定义一个用于存储用户信息的实体对象,代码如下:

```java
package com.lietu.shiro.domain;
public class UserDO {
    private Integer id;
    private String userName;
    private String password;
    public Integer getId() {
        return id;
    }
    public void setId(Integer id) {
        this.id = id;
    }
    public String getUserName() {
```

```
        return userName;
    }
    public void setUserName(String userName) {
        this.userName = userName;
    }
    public String getPassword() {
        return password;
    }
    public void setPassword(String password) {
        this.password = password;
    }
}
```

UserRealm 类实现自定义认证与授权，代码如下：

```
package com.lietu.shiro.config;
import com.lietu.shiro.domain.UserDO;
import org.apache.shiro.authc.*;
import org.apache.shiro.authz.AuthorizationInfo;
import org.apache.shiro.authz.SimpleAuthorizationInfo;
import org.apache.shiro.realm.AuthorizingRealm;
import org.apache.shiro.subject.PrincipalCollection;
import java.util.*;

/**
 * 重写授权和认证的方法
 **/
public class UserRealm extends AuthorizingRealm {
    /**
     * 重写认证
     * @param authenticationToken token
     * @return 返回认证信息实体
     **/
    @Override
    protected AuthenticationInfo doGetAuthenticationInfo(AuthenticationToken authenticationToken) throws AuthenticationException {
        String username=(String)authenticationToken.getPrincipal();//身份,例如用户名
        Map<String ,Object> map=new HashMap<>(16);
        map.put("username",username);
        String password=new String((char[]) authenticationToken.getCredentials());
        //证明,例如密码
        //对身份+证明的数据认证,这里模拟了一个数据源
        //如果是数据库,那么这里应该调用数据库判断用户名及密码是否正确
        if(!"admin".equals(username) || !"123456".equals(password)){
            throw new IncorrectCredentialsException("账号或密码不正确");
        }
        //认证通过
        UserDO user=new UserDO();
        user.setId(1);//假设用户ID=1
        user.setUserName(username);
        user.setPassword(password);
        //建立一个SimpleAuthenticationInfo认证模块,包括身份、证明等信息
        SimpleAuthenticationInfo info = new SimpleAuthenticationInfo(user, password,
```

```
getName());
        return info;
    }

    /**
     * 重写授权
     * @param principalCollection 身份信息
     * @return 返回授权信息对象
     */
    @Override
    protected AuthorizationInfo doGetAuthorizationInfo(PrincipalCollection principalCollection) {
        UserDO userDO = (UserDO)principalCollection.getPrimaryPrincipal();
        Integer userId= userDO.getId();//转成 user 对象
        //新建一个授权模块 SimpleAuthorizationInfo,把权限赋给当前的用户
        SimpleAuthorizationInfo info = new SimpleAuthorizationInfo();

        //设置当前会话可拥有的角色,实际场景根据业务来获取角色列表（如从数据库获取角色列表）
        Set<String> roles=new HashSet<>();
        roles.add("admin");
        roles.add("finance");
        info.setRoles(roles);

        //设置当前会话可拥有的权限,实际场景根据业务来获取权限列表（如从数据库获取角色列表下的权限列表）
        Set<String> permissions=new HashSet<>();
        permissions.add("system:article:article");
        permissions.add("system:article:add");
        permissions.add("system:article:edit");
        permissions.add("system:article:remove");
        permissions.add("system:article:batchRemove");
        info.setStringPermissions(permissions);
        return  info;
    }
}
```

配置类 ShiroConfig 的实现代码如下：

```
package com.lietu.shiro.config;
import at.pollux.thymeleaf.shiro.dialect.ShiroDialect;
import org.apache.shiro.spring.LifecycleBeanPostProcessor;
import org.apache.shiro.spring.security.interceptor.AuthorizationAttributeSourceAdvisor;
import org.apache.shiro.spring.web.ShiroFilterFactoryBean;
import org.apache.shiro.web.mgt.DefaultWebSecurityManager;
import org.springframework.aop.framework.autoproxy.DefaultAdvisorAutoProxyCreator;
import org.springframework.context.annotation.Bean;
import org.springframework.context.annotation.Configuration;
import org.apache.shiro.mgt.SecurityManager;
import java.util.LinkedHashMap;

@Configuration
public class ShiroConfig {
    @Bean
    public static LifecycleBeanPostProcessor getLifecycleBeanPostProcessor() {
```

```java
        return new LifecycleBeanPostProcessor();
    }

    /**
     * 开启shiro aop注解支持,如@RequiresRoles、@RequiresPermissions
     * 使用代理方式,所以需要开启代码支持
     * @param securityManager
     * @return
     */
    @Bean
    public AuthorizationAttributeSourceAdvisor authorizationAttributeSourceAdvisor(SecurityManager securityManager){
        AuthorizationAttributeSourceAdvisor authorizationAttributeSourceAdvisor =
                new AuthorizationAttributeSourceAdvisor();
        authorizationAttributeSourceAdvisor.setSecurityManager(securityManager);
        return authorizationAttributeSourceAdvisor;
    }

    /**
     * 开启shiro aop注解支持
     * */
    @Bean
    public DefaultAdvisorAutoProxyCreator advisorAutoProxyCreator(){
        DefaultAdvisorAutoProxyCreator advisorAutoProxyCreator =
                new DefaultAdvisorAutoProxyCreator();
        advisorAutoProxyCreator.setProxyTargetClass(true);
        return advisorAutoProxyCreator;
    }

    /**
     * ShiroDialect,是为了在thymeleaf中使用shiro标签的Bean而设置的
     * @return
     */
    @Bean
    public ShiroDialect shiroDialect() {
        return new ShiroDialect();
    }

    /**
     * shiroFilterFactoryBean 实现过滤器过滤
     * setFilterChainDefinitionMap 设置可访问或禁止访问的目录
     * @param securityManager 表示安全管理器
     * */
    @Bean
    ShiroFilterFactoryBean shiroFilterFactoryBean(SecurityManager securityManager) {
        ShiroFilterFactoryBean shiroFilterFactoryBean = new ShiroFilterFactoryBean();
        shiroFilterFactoryBean.setSecurityManager(securityManager);
        //设置登录页面
        shiroFilterFactoryBean.setLoginUrl("/login");
        //登录后的页面
        shiroFilterFactoryBean.setSuccessUrl("/index");
        //未认证页面提示
```

```
            shiroFilterFactoryBean.setUnauthorizedUrl("/403");
            //设置无须加载权限的页面过滤器
            LinkedHashMap<String, String> filterChainDefinitionMap = new LinkedHashMap<>();
            filterChainDefinitionMap.put("/fonts/**", "anon");
            filterChainDefinitionMap.put("/css/**", "anon");
            filterChainDefinitionMap.put("/js/**", "anon");
            //authc 表示有权限;anon 表示不需要权限;** 表示某个目录下所有的文件
            //filterChainDefinitionMap.put("/**", "anon"); //设置所有页面不需要登录即可见
                                                           //一般用于某个文件夹下方网站
            filterChainDefinitionMap.put("/**", "authc");  //设置所有页面都需要登录才可见
            //设置过滤器
            shiroFilterFactoryBean.setFilterChainDefinitionMap(filterChainDefinitionMap);
            return shiroFilterFactoryBean;
        }
        /**
         * 获取用户令牌,这里的令牌相当于你去公园时手里的门票。"门票"只有放在安全系统中才能识别,这里是
指放在 SecurityManager 中
         * */
        @Bean
        UserRealm userRealm() {
            UserRealm userRealm = new UserRealm();
            return userRealm;
        }

        @Bean
        public SecurityManager securityManager() {
            DefaultWebSecurityManager securityManager = new DefaultWebSecurityManager();
            // 设置 realm
            securityManager.setRealm(userRealm());
            return securityManager;
        }
    }
```

添加登录页面 resources/templates/login.html，代码如下：

```
<!DOCTYPE html>
<html lang="en">
    <head>
        <meta charset="UTF-8">
        <title>使用 shiro 登录页面</title>
    </head>
    <body>
    <div>
        <input id="userName" name="userName" value="">
    </div>
    <div>
        <input id="password" name="password" value="">
    </div>
    <div>
        <input type="button" id="btnSave" value="登录">
    </div>
    <script src="https://cdn.bootcss.com/jquery/1.11.3/jquery.js"></script>
    <script>
```

```javascript
$(function() {
    $("#btnSave").click(function () {
        var username=$("#userName").val();
        var password=$("#password").val();
        $.ajax({
            cache: true,
            type: "POST",
            url: "/login",
            data: "userName=" + username + "&password=" + password,
            dataType: "json",
            async: false,
            error: function (request) {
                console.log("Connection error");
            },
            success: function (data) {
                if (data.status == 0) {
                    window.location = "/index";
                    return false;
                } else {
                    alert(data.message);
                }
            }
        });
    });
});
</script>
</body>
</html>
```

控制类 UserController 的实现代码如下：

```java
package com.lietu.shiro.controller;
import org.apache.shiro.SecurityUtils;
import org.apache.shiro.authc.AuthenticationException;
import org.apache.shiro.authc.UsernamePasswordToken;
import org.apache.shiro.authz.UnauthenticatedException;
import org.apache.shiro.authz.annotation.RequiresPermissions;
import org.apache.shiro.subject.Subject;
import org.springframework.stereotype.Controller;
import org.springframework.ui.Model;
import org.springframework.web.bind.annotation.*;
import org.springframework.web.bind.support.SessionStatus;
import javax.servlet.http.HttpSession;
import java.util.HashMap;
import java.util.Map;

@Controller
public class UserController {
    //shiro 认证成功后默认跳转页面
    @GetMapping("/index")
    public String index(){
        return "index";
    }
```

```java
@GetMapping("/403")
public String err403(){
    return "403";
}
/**
 * 根据权限授权使用注解 @RequiresPermissions
 * */
@GetMapping("/article")
@RequiresPermissions("system:article:article")
public String article(){
    return "article";
}

/**
 * 根据权限授权使用注解 @RequiresPermissions
 * */
@GetMapping("/setting")
@RequiresPermissions("system:setting:setting")
public String setting(){
    return "setting";
}
@GetMapping("/show")
@ResponseBody
public String show(){
    Subject subject = SecurityUtils.getSubject();
    String str="";
    if(subject.hasRole("admin")){
        str=str+"您拥有admin权限";
    }else{
        str=str+"您没有admin权限";
    }
    if(subject.hasRole("sale")){
        str=str+"您拥有sale权限";
    }
    else{
        str=str+"您没有sale权限";
    }
    try{
        subject.checkPermission("app:setting:setting");
        str=str+"您拥有system:setting:setting权限";

    }catch (UnauthenticatedException ex){
        str=str+"您没有system:setting:setting权限";
    }
    return  str;
}

//get /login方法,对应前端 login.html 页面
@GetMapping("/login")
public String login(){
    return "login";
}
```

```java
//post /login 方法,对应登录提交接口
@PostMapping("/login")
@ResponseBody
public Object loginsubmit(@RequestParam String userName,@RequestParam String password){
    Map<String,Object> map=new HashMap<>();
    //把身份 useName 和证明 password 封装成对象 UsernamePasswordToken
    UsernamePasswordToken token=new UsernamePasswordToken(userName,password);
    //获取当前的 subject
    //Subject 可以是一个用户,也有可能是一个第三方程序
    Subject subject = SecurityUtils.getSubject();
    try{
        subject.login(token);
        map.put("status",0);
        map.put("message","登录成功");
        return map;
    }catch (AuthenticationException e){
        map.put("status",1);
        map.put("message","用户名或密码错误");
        return map;
    }
}

@GetMapping("/logout")
String logout(HttpSession session, SessionStatus sessionStatus, Model model) {
    //会员中心退出登录
    session.removeAttribute("userData");
    sessionStatus.setComplete();
    SecurityUtils.getSubject().logout();         //调用 subject 的 logout 方法进行注销
    return "redirect:/login";
}
}
```

登录成功后跳转到的页面 resources/templates/index.html,代码如下:

```html
<!DOCTYPE html>
<html lang="en">
  <head>
    <meta charset="UTF-8">
    <title>通过登录验证后跳转到此页面</title>
  </head>
  <body>
    <div>
        <a href="/article">前往文章页面</a>
    </div>
    <div>
        <a href="/setting">前往设置页面</a>
    </div>
  </body>
</html>
```

已授权访问的页面 resources/templates/article.html,代码如下:

```html
<!DOCTYPE html>
```

```html
<html lang="en">
  <head>
    <meta charset="UTF-8">
    <title>必须获取 system:article 授权</title>
  </head>
  <body>
//必须获取 system:article 授权才会显示
  </body>
</html>
```

未授权访问的页面 resources/templates/setting.html，代码如下：

```html
<!DOCTYPE html>
<html lang="en">
  <head>
    <meta charset="UTF-8">
    <title>必须获取 system:setting 授权</title>
  </head>
  <body>
//必须获取 system:setting 授权才会显示
  </body>
</html>
```

未授权统一页面 resources/templates/error/403.html，代码如下：

```html
<!DOCTYPE html>
<html lang="en">
  <head>
    <meta charset="UTF-8">
    <title>403 没有授权</title>
  </head>
  <body>
//你访问的页面没有授权
  </body>
</html>
```

4.3 集成 JWT 身份验证

如今，JWT（JSON Web Token）在身份验证和信息交换方面应用很流行。服务器不创建会话（基于会话的身份验证），而是将数据编码为 JSON Web 令牌并将其发送给客户端。客户端保存 JWT，然后从客户端到受保护路由或资源的每个请求都应附加该 JWT（通常在标头处）。服务器将验证该 JWT 并返回响应。

Bearer 类型的令牌看起来像以下这样：

```
$curl -H "Authorization: Bearer <ACCESS_TOKEN>" http://www.example.com
```

与需要将会话存储在 Cookie 上的基于会话的身份验证相比，JWT（基于令牌的身份验证）的最大优势在于，将令牌存储在客户端：浏览器的本地存储、iOS 的 keyChain（钥匙串）和 Android 的 SharedPreferences 等，所以无须构建另一个支持本机应用程序的后端项目或针对本机应用程序用户的其他身份验证模块。

构建一个 Spring Boot 应用程序，需要注意以下几点。
- 用户可以注册新账户，或使用已拥有的用户名和密码登录。
- 根据用户的角色（管理员、主持人、用户）来授权用户访问资源。3 个角色对应于 3 个枚举值，代码如下：

```
package com.bezkoder.springjwt.models;
public enum ERole {
    ROLE_USER,
    ROLE_MODERATOR,
    ROLE_ADMIN
}
```

打开 pom.xml 并添加以下依赖项。

```
<dependency>
    <groupId>org.springframework.boot</groupId>
    <artifactId>spring-boot-starter-data-jpa</artifactId>
</dependency>
<dependency>
    <groupId>org.springframework.boot</groupId>
    <artifactId>spring-boot-starter-security</artifactId>
</dependency>
<dependency>
    <groupId>org.springframework.boot</groupId>
    <artifactId>spring-boot-starter-web</artifactId>
</dependency>
<dependency>
    <groupId>io.jsonwebtoken</groupId>
    <artifactId>jjwt</artifactId>
    <version>0.9.1</version>
</dependency>
<dependency>
    <groupId>mysql</groupId>
    <artifactId>mysql-connector-java</artifactId>
    <scope>runtime</scope>
</dependency>
```

配置 Spring 数据源，代码如下：

```
spring.datasource.url= jdbc:mysql://localhost:3306/testdb?useSSL=false
spring.datasource.username= root
spring.datasource.password= 123456
spring.jpa.properties.hibernate.dialect= org.hibernate.dialect.MySQL5InnoDBDialect
spring.jpa.hibernate.ddl-auto= update

# App Properties
bezkoder.app.jwtSecret= bezKoderSecretKey
bezkoder.app.jwtExpirationMs= 86400000
```

数据库中有 3 个表：users、roles 和用于多对多关系的 user_roles。下面介绍相应角色模型和用户模型的实现代码。

Role.java 中的角色模型实现代码如下：

```
package com.bezkoder.springjwt.models;
```

```java
import javax.persistence.*;

@Entity
@Table(name = "roles")
public class Role {
    @Id
    @GeneratedValue(strategy = GenerationType.IDENTITY)
    private Integer id;
    @Enumerated(EnumType.STRING)
    @Column(length = 20)
    private ERole name;
    public Role() {

    }
    public Role(ERole name) {
        this.name = name;
    }
    public Integer getId() {
        return id;
    }
    public void setId(Integer id) {
        this.id = id;
    }
    public ERole getName() {
        return name;
    }
    public void setName(ERole name) {
        this.name = name;
    }
}
```

User.java 中的用户模型实现代码如下：

```java
package com.bezkoder.springjwt.models;
import java.util.HashSet;
import java.util.Set;
import javax.persistence.*;
import javax.validation.constraints.Email;
import javax.validation.constraints.NotBlank;
import javax.validation.constraints.Size;

@Entity
@Table(   name = "users",
      uniqueConstraints = {
         @UniqueConstraint(columnNames = "username"),
         @UniqueConstraint(columnNames = "email")
      })
public class User {
    @Id
    @GeneratedValue(strategy = GenerationType.IDENTITY)
    private Long id;

    @NotBlank
```

```java
    @Size(max = 20)
    private String username;

    @NotBlank
    @Size(max = 50)
    @Email
    private String email;

    @NotBlank
    @Size(max = 120)
    private String password;

    @ManyToMany(fetch = FetchType.LAZY)
    @JoinTable(   name = "user_roles",
            joinColumns = @JoinColumn(name = "user_id"),
            inverseJoinColumns = @JoinColumn(name = "role_id"))
    private Set<Role> roles = new HashSet<>();
    public User() {
    }
    public User(String username, String email, String password) {
        this.username = username;
        this.email = email;
        this.password = password;
    }

    public Long getId() {
        return id;
    }

    public void setId(Long id) {
        this.id = id;
    }

    public String getUsername() {
        return username;
    }

    public void setUsername(String username) {
        this.username = username;
    }

    public String getEmail() {
        return email;
    }

    public void setEmail(String email) {
        this.email = email;
    }

    public String getPassword() {
        return password;
    }
```

```java
    public void setPassword(String password) {
        this.password = password;
    }
    public Set<Role> getRoles() {
        return roles;
    }
    public void setRoles(Set<Role> roles) {
        this.roles = roles;
    }
}
```

以上每个模型都需要一个用于持久存储和访问数据的存储库。在存储库包中，创建两个存储库。用户存储库的实现代码如下：

```java
package com.bezkoder.springjwt.repository;
import java.util.Optional;
import org.springframework.data.jpa.repository.JpaRepository;
import org.springframework.stereotype.Repository;
import com.bezkoder.springjwt.models.User;

@Repository
public interface UserRepository extends JpaRepository<User, Long> {
    Optional<User> findByUsername(String username);
    Boolean existsByUsername(String username);
    Boolean existsByEmail(String email);
}
```

角色存储库扩展了 JpaRepository，实现代码如下：

```java
package com.bezkoder.springjwt.repository;
import java.util.Optional;
import org.springframework.data.jpa.repository.JpaRepository;
import org.springframework.stereotype.Repository;
import com.bezkoder.springjwt.models.ERole;
import com.bezkoder.springjwt.models.Role;

@Repository
public interface RoleRepository extends JpaRepository<Role, Long> {
    Optional<Role> findByName(ERole name);
}
```

WebSecurityConfig 类扩展了 WebSecurityConfigurerAdapter，实现代码如下：

```java
package com.bezkoder.springjwt.security;
import org.springframework.beans.factory.annotation.Autowired;
import org.springframework.context.annotation.Bean;
import org.springframework.context.annotation.Configuration;
import org.springframework.security.authentication.AuthenticationManager;
import org.springframework.security.config.annotation.authentication.builders.AuthenticationManagerBuilder;
import org.springframework.security.config.annotation.method.configuration.EnableGlobalMethodSecurity;
import org.springframework.security.config.annotation.web.builders.HttpSecurity;
import org.springframework.security.config.annotation.web.configuration.EnableWeb
```

```java
Security;
import org.springframework.security.config.annotation.web.configuration.WebSecurityConfigurerAdapter;
import org.springframework.security.config.http.SessionCreationPolicy;
import org.springframework.security.crypto.bcrypt.BCryptPasswordEncoder;
import org.springframework.security.crypto.password.PasswordEncoder;
import org.springframework.security.web.authentication.UsernamePasswordAuthenticationFilter;
import com.bezkoder.springjwt.security.jwt.AuthEntryPointJwt;
import com.bezkoder.springjwt.security.jwt.AuthTokenFilter;
import com.bezkoder.springjwt.security.services.UserDetailsServiceImpl;

@Configuration
@EnableWebSecurity
@EnableGlobalMethodSecurity(
    // securedEnabled = true,
    // jsr250Enabled = true,
    prePostEnabled = true)
public class WebSecurityConfig extends WebSecurityConfigurerAdapter {
    @Autowired
    UserDetailsServiceImpl userDetailsService;
    @Autowired
    private AuthEntryPointJwt unauthorizedHandler;
    @Bean
    public AuthTokenFilter authenticationJwtTokenFilter() {
        return new AuthTokenFilter();
    }
    @Override
    public void configure(AuthenticationManagerBuilder authenticationManagerBuilder) throws Exception {
        authenticationManagerBuilder.userDetailsService(userDetailsService).passwordEncoder(passwordEncoder());
    }

    @Bean
    @Override
    public AuthenticationManager authenticationManagerBean() throws Exception {
        return super.authenticationManagerBean();
    }

    @Bean
    public PasswordEncoder passwordEncoder() {
        return new BCryptPasswordEncoder();
    }

    @Override
    protected void configure(HttpSecurity http) throws Exception {
        http.cors().and().csrf().disable()
            .exceptionHandling().authenticationEntryPoint(unauthorizedHandler).and()
            .sessionManagement().sessionCreationPolicy(SessionCreationPolicy.STATELESS).and()
            .authorizeRequests().antMatchers("/api/auth/**").permitAll()
```

```
                .antMatchers("/api/test/**").permitAll()
                .anyRequest().authenticated();

        http.addFilterBefore(authenticationJwtTokenFilter(), UsernamePassword
AuthenticationFilter.class);
    }
}
```

在上面的代码实现过程中，需要注意以下几点。
- @EnableWebSecurity 允许 Spring Boot 查找该类并将其自动应用于全局 Web Security。
- @EnableGlobalMethodSecurity 在方法上提供 AOP 安全性。它启用@PreAuthorize、@PostAuthorize，还支持 JSR-250。此外，还可以在方法安全性表达式的配置中找到更多参数。
- 从 WebSecurityConfigurerAdapter 接口重写 configure 方法。它告知 Spring Security 如何配置 CORS（Cross-origin Resource Sharing，跨域资源共享）和 CSRF（Cross-site Request Forgery，跨站请求伪造）、何时要求对所有用户进行身份验证或者不验证、何时使用哪个过滤器（AuthTokenFilter）及何时使它工作（在 UsernamePasswordAuthenticationFilter 之前的过滤器）、选择哪个异常处理程序（AuthEntryPointJwt）。
- Spring Security 将加载用户详细信息以执行认证和授权。因此，它具有我们需要实现的 UserDetailsService 接口。UserDetailsService 的实现将用于通过 AuthenticationManagerBuilder.userDetailsService()方法配置 DaoAuthenticationProvider。另外，还需要为 DaoAuthenticationProvider 提供 PasswordEncoder。如果未指定，DaoAuthenticationProvider 将使用纯文本。

如果验证过程成功，可以从 Authentication 对象获取用户信息，例如用户名、密码、权限，代码如下：

```
Authentication authentication =
        authenticationManager.authenticate(
            new UsernamePasswordAuthenticationToken(username, password)
        );
UserDetails userDetails = (UserDetails) authentication.getPrincipal();
// userDetails.getUsername()
// userDetails.getPassword()
// userDetails.getAuthorities()
```

如果我们想获取更多数据（标识、电子邮件等），则可以创建此 UserDetails 接口的实现，代码如下：

```
package com.bezkoder.springjwt.security.services;
import java.util.Collection;
import java.util.List;
import java.util.Objects;
import java.util.stream.Collectors;
import org.springframework.security.core.GrantedAuthority;
import org.springframework.security.core.authority.SimpleGrantedAuthority;
import org.springframework.security.core.userdetails.UserDetails;
import com.bezkoder.springjwt.models.User;
```

```java
import com.fasterxml.jackson.annotation.JsonIgnore;

public class UserDetailsImpl implements UserDetails {
    private static final long serialVersionUID = 1L;
    private Long id;
    private String username;
    private String email;
    @JsonIgnore
    private String password;
    private Collection<? extends GrantedAuthority> authorities;
    public UserDetailsImpl(Long id, String username, String email, String password,
            Collection<? extends GrantedAuthority> authorities) {
        this.id = id;
        this.username = username;
        this.email = email;
        this.password = password;
        this.authorities = authorities;
    }
    public static UserDetailsImpl build(User user) {
        List<GrantedAuthority> authorities = user.getRoles().stream()
                .map(role -> new SimpleGrantedAuthority(role.getName().name()))
                .collect(Collectors.toList());
        return new UserDetailsImpl(
                user.getId(),
                user.getUsername(),
                user.getEmail(),
                user.getPassword(),
                authorities);
    }
    @Override
    public Collection<? extends GrantedAuthority> getAuthorities() {
        return authorities;
    }
    public Long getId() {
        return id;
    }
    public String getEmail() {
        return email;
    }
    @Override
    public String getPassword() {
        return password;
    }
    @Override
    public String getUsername() {
        return username;
    }
    @Override
    public boolean isAccountNonExpired() {
        return true;
    }
    @Override
```

```java
    public boolean isAccountNonLocked() {
        return true;
    }
    @Override
    public boolean isCredentialsNonExpired() {
        return true;
    }
    @Override
    public boolean isEnabled() {
        return true;
    }
    @Override
    public boolean equals(Object o) {
        if (this == o)
            return true;
        if (o == null || getClass() != o.getClass())
            return false;
        UserDetailsImpl user = (UserDetailsImpl) o;
        return Objects.equals(id, user.id);
    }
}
```

需要通过 UserDetailsService 来获取 UserDetails 对象。从以下代码中可以看到,UserDetailsService 接口只有一个方法。

```java
public interface UserDetailsService {
    UserDetails loadUserByUsername(String username) throws UsernameNotFoundException;
}
```

下面编程实现以上需求并覆盖 loadUserByUsername()方法。

```java
package com.bezkoder.springjwt.security.services;
import org.springframework.beans.factory.annotation.Autowired;
import org.springframework.security.core.userdetails.UserDetails;
import org.springframework.security.core.userdetails.UserDetailsService;
import org.springframework.security.core.userdetails.UsernameNotFoundException;
import org.springframework.stereotype.Service;
import org.springframework.transaction.annotation.Transactional;
import com.bezkoder.springjwt.models.User;
import com.bezkoder.springjwt.repository.UserRepository;

@Service
public class UserDetailsServiceImpl implements UserDetailsService {
    @Autowired
    UserRepository userRepository;

    @Override
    @Transactional
    public UserDetails loadUserByUsername(String username) throws UsernameNotFoundException {
        User user = userRepository.findByUsername(username)
                .orElseThrow(() -> new UsernameNotFoundException("User Not Found with username: " + username));
        return UserDetailsImpl.build(user);
```

        }
    }

在上面的代码中,使用 UserRepository 获得了完整的自定义 User 对象,然后使用静态 build() 方法构建了 UserDetails 对象。

接下来,创建一个过滤器,该过滤器针对每个请求都执行一次。创建过滤器 AuthTokenFilter 类,该类扩展了 OncePerRequestFilter 并覆盖了 doFilterInternal()方法,代码如下:

```
package com.bezkoder.springjwt.security.jwt;
import java.io.IOException;
import javax.servlet.FilterChain;
import javax.servlet.ServletException;
import javax.servlet.http.HttpServletRequest;
import javax.servlet.http.HttpServletResponse;
import org.slf4j.Logger;
import org.slf4j.LoggerFactory;
import org.springframework.beans.factory.annotation.Autowired;
import org.springframework.security.authentication.UsernamePasswordAuthenticationToken;
import org.springframework.security.core.context.SecurityContextHolder;
import org.springframework.security.core.userdetails.UserDetails;
import org.springframework.security.web.authentication.WebAuthenticationDetailsSource;
import org.springframework.util.StringUtils;
import org.springframework.web.filter.OncePerRequestFilter;
import com.bezkoder.springjwt.security.services.UserDetailsServiceImpl;

public class AuthTokenFilter extends OncePerRequestFilter {
    @Autowired
    private JwtUtils jwtUtils;
    @Autowired
    private UserDetailsServiceImpl userDetailsService;
    private static final Logger logger = LoggerFactory.getLogger(AuthTokenFilter.class);

    @Override
    protected void doFilterInternal(HttpServletRequest request, HttpServletResponse response, FilterChain filterChain) throws ServletException, IOException {
        try {
            String jwt = parseJwt(request);
            if (jwt != null && jwtUtils.validateJwtToken(jwt)) {
                String username = jwtUtils.getUserNameFromJwtToken(jwt);
                UserDetails userDetails = userDetailsService.loadUserByUsername(username);
                UsernamePasswordAuthenticationToken authentication = new UsernamePasswordAuthenticationToken(userDetails, null, userDetails.getAuthorities());
                authentication.setDetails(new WebAuthenticationDetailsSource().buildDetails(request));
                SecurityContextHolder.getContext().setAuthentication(authentication);
            }
        } catch (Exception e) {
            logger.error("Cannot set user authentication: {}", e);
        }
        filterChain.doFilter(request, response);
    }
```

```java
    private String parseJwt(HttpServletRequest request) {
        String headerAuth = request.getHeader("Authorization");
        if (StringUtils.hasText(headerAuth) && headerAuth.startsWith("Bearer ")) {
            return headerAuth.substring(7, headerAuth.length());
        }
        return null;
    }
}
```

在以上代码中，doFilterInternal()方法内部进行了如下操作。

①从Authorization标头中获取JWT（通过删除Bearer前缀）。

②如果请求中包含JWT，对其进行验证，然后从中解析用户名。

③从用户名中获取UserDetails，以创建一个Authentication对象。

④使用setAuthentication(authentication)方法在SecurityContext中设置当前的UserDetails。

创建Jwt Utils类，此类具有以下3个功能。

- 根据用户名、日期、有效期和密码生成JWT。
- 从JWT获取用户名。
- 验证JWT。

JwtUtils类的实现代码如下：

```java
package com.bezkoder.springjwt.security.jwt;
import java.util.Date;
import org.slf4j.Logger;
import org.slf4j.LoggerFactory;
import org.springframework.beans.factory.annotation.Value;
import org.springframework.security.core.Authentication;
import org.springframework.stereotype.Component;
import com.bezkoder.springjwt.security.services.UserDetailsImpl;
import io.jsonwebtoken.*;

@Component
public class JwtUtils {
    private static final Logger logger = LoggerFactory.getLogger(JwtUtils.class);
    @Value("${bezkoder.app.jwtSecret}")
    private String jwtSecret;
    @Value("${bezkoder.app.jwtExpirationMs}")
    private int jwtExpirationMs;

    public String generateJwtToken(Authentication authentication) {
        UserDetailsImpl userPrincipal = (UserDetailsImpl) authentication.getPrincipal();
        return Jwts.builder()
                .setSubject((userPrincipal.getUsername()))
                .setIssuedAt(new Date())
                .setExpiration(new Date((new Date()).getTime() + jwtExpirationMs))
                .signWith(SignatureAlgorithm.HS512, jwtSecret)
                .compact();
    }
    public String getUserNameFromJwtToken(String token) {
```

```java
            return Jwts.parser().setSigningKey(jwtSecret).parseClaimsJws(token).getBody().getSubject();
    }
    public boolean validateJwtToken(String authToken) {
        try {
            Jwts.parser().setSigningKey(jwtSecret).parseClaimsJws(authToken);
            return true;
        } catch (SignatureException e) {
            logger.error("Invalid JWT signature: {}", e.getMessage());
        } catch (MalformedJwtException e) {
            logger.error("Invalid JWT token: {}", e.getMessage());
        } catch (ExpiredJwtException e) {
            logger.error("JWT token is expired: {}", e.getMessage());
        } catch (UnsupportedJwtException e) {
            logger.error("JWT token is unsupported: {}", e.getMessage());
        } catch (IllegalArgumentException e) {
            logger.error("JWT claims string is empty: {}", e.getMessage());
        }

        return false;
    }
}
```

创建实现 AuthenticationEntryPoint 接口的 AuthEntryPointJwt 类，并重写 commence()方法。每当未经身份验证的用户请求安全的 HTTP 资源并抛出 AuthenticationException 时，都会触发此方法。

```java
package com.bezkoder.springjwt.security.jwt;
import java.io.IOException;
import javax.servlet.ServletException;
import javax.servlet.http.HttpServletRequest;
import javax.servlet.http.HttpServletResponse;
import org.slf4j.Logger;
import org.slf4j.LoggerFactory;
import org.springframework.security.core.AuthenticationException;
import org.springframework.security.web.AuthenticationEntryPoint;
import org.springframework.stereotype.Component;

@Component
public class AuthEntryPointJwt implements AuthenticationEntryPoint {
    private static final Logger logger = LoggerFactory.getLogger(AuthEntryPointJwt.class);

    @Override
    public void commence(HttpServletRequest request, HttpServletResponse response, AuthenticationException authException) throws IOException, ServletException {
        logger.error("Unauthorized error: {}", authException.getMessage());
        response.sendError(HttpServletResponse.SC_UNAUTHORIZED, "Error: Unauthorized");
    }
}
```

在以上代码中，HttpServletResponse.SC_UNAUTHORIZED 的状态码为 401，它表示该请求需要 HTTP 身份验证。

至此，已经为 Spring Security 构建了所有部分。接下来，将展示如何为 RESTful API 实现控制器。

请求 RESTful API 的有效负载包括以下两个。
- LoginRequest: { username, password }。
- SignupRequest: { username, email, password }。

响应 RESTful API 的有效负载包括以下两个。
- JwtResponse: { token, type, id, username, email, roles }。
- MessageResponse: { message }。

用于认证的控制器提供用于注册和登录操作的 API。其中，
- /api/auth/signup：
  - 检查现有的用户名/电子邮件；
  - 创建新用户（如果未指定角色，则使用 ROLE_USER）；
  - 使用 UserRepository 将用户信息保存到数据库。
- /api/auth/signin：
  - 验证用户名、密码；
  - 使用身份验证对象更新 SecurityContext；
  - 生成 JWT；
  - 从身份验证对象获取 UserDetails。

controllers/AuthController.java 实现代码如下：

```
package com.bezkoder.springjwt.controllers;
import java.util.HashSet;
import java.util.List;
import java.util.Set;
import java.util.stream.Collectors;
import javax.validation.Valid;
import org.springframework.beans.factory.annotation.Autowired;
import org.springframework.http.ResponseEntity;
import org.springframework.security.authentication.AuthenticationManager;
import org.springframework.security.authentication.UsernamePasswordAuthenticationToken;
import org.springframework.security.core.Authentication;
import org.springframework.security.core.context.SecurityContextHolder;
import org.springframework.security.crypto.password.PasswordEncoder;
import org.springframework.web.bind.annotation.CrossOrigin;
import org.springframework.web.bind.annotation.PostMapping;
import org.springframework.web.bind.annotation.RequestBody;
import org.springframework.web.bind.annotation.RequestMapping;
import org.springframework.web.bind.annotation.RestController;
import com.bezkoder.springjwt.models.ERole;
import com.bezkoder.springjwt.models.Role;
import com.bezkoder.springjwt.models.User;
import com.bezkoder.springjwt.payload.request.LoginRequest;
import com.bezkoder.springjwt.payload.request.SignupRequest;
import com.bezkoder.springjwt.payload.response.JwtResponse;
import com.bezkoder.springjwt.payload.response.MessageResponse;
```

```java
    import com.bezkoder.springjwt.repository.RoleRepository;
    import com.bezkoder.springjwt.repository.UserRepository;
    import com.bezkoder.springjwt.security.jwt.JwtUtils;
    import com.bezkoder.springjwt.security.services.UserDetailsImpl;
    @CrossOrigin(origins = "*", maxAge = 3600)
    @RestController
    @RequestMapping("/api/auth")
    public class AuthController {
        @Autowired
        AuthenticationManager authenticationManager;
        @Autowired
        UserRepository userRepository;
        @Autowired
        RoleRepository roleRepository;
        @Autowired
        PasswordEncoder encoder;
        @Autowired
        JwtUtils jwtUtils;
        @PostMapping("/signin")
        public ResponseEntity<?> authenticateUser(@Valid @RequestBody LoginRequest loginRequest) {
            Authentication authentication = authenticationManager.authenticate(
    new UsernamePasswordAuthenticationToken(loginRequest.getUsername(), loginRequest.getPassword()));
            SecurityContextHolder.getContext().setAuthentication(authentication);
            String jwt = jwtUtils.generateJwtToken(authentication);
            UserDetailsImpl userDetails = (UserDetailsImpl) authentication.getPrincipal();
               List<String> roles = userDetails.getAuthorities().stream()
                    .map(item -> item.getAuthority())
                    .collect(Collectors.toList());
            return ResponseEntity.ok(new JwtResponse(jwt,
                                                 userDetails.getId(),
                                                 userDetails.getUsername(),
                                                 userDetails.getEmail(),
                                                 roles));
        }

        @PostMapping("/signup")
        public ResponseEntity<?> registerUser(@Valid @RequestBody SignupRequest signUpRequest) {
            if (userRepository.existsByUsername(signUpRequest.getUsername())) {
                return ResponseEntity
                    .badRequest()
                    .body(new MessageResponse("Error: Username is already taken!"));
            }
            if (userRepository.existsByEmail(signUpRequest.getEmail())) {
                return ResponseEntity
                    .badRequest()
                    .body(new MessageResponse("Error: Email is already in use!"));
            }

            // Create new user's account
```

```java
            User user = new User(signUpRequest.getUsername(),
                         signUpRequest.getEmail(),
                         encoder.encode(signUpRequest.getPassword()));
        Set<String> strRoles = signUpRequest.getRole();
        Set<Role> roles = new HashSet<>();
        if (strRoles == null) {
            Role userRole = roleRepository.findByName(ERole.ROLE_USER)
                .orElseThrow(() -> new RuntimeException("Error: Role is not found."));
            roles.add(userRole);
        } else {
            strRoles.forEach(role -> {
                switch (role) {
                case "admin":
                    Role adminRole = roleRepository.findByName(ERole.ROLE_ADMIN).orElseThrow(() -> new RuntimeException("Error: Role is not found."));
                    roles.add(adminRole);
                    break;
                case "mod":
                    Role modRole = roleRepository.findByName(ERole.ROLE_MODERATOR).orElseThrow(() -> new RuntimeException("Error: Role is not found."));
                    roles.add(modRole);
                    break;
                default:
                    Role userRole = roleRepository.findByName(ERole.ROLE_USER)
                        .orElseThrow(() -> new RuntimeException("Error: Role is not found."));
                    roles.add(userRole);
                }
            });
        }
        user.setRoles(roles);
        userRepository.save(user);
        return ResponseEntity.ok(new MessageResponse("User registered successfully!"));
    }
}
```

用于测试授权的控制器有以下 4 个 API。其中，

- /api/test/all 供公开访问；
- /api/test/user 对应具有 ROLE_USER、ROLE_MODERATOR 或 ROLE_ADMIN 的用户；
- /api/test/mod 对应具有 ROLE_MODERATOR 的用户；
- /api/test/admin 对应具有 ROLE_ADMIN 的用户。

可以使用@PreAuthorize 注解轻松地在 Apis 中保护方法。

controllers/TestController.java 的实现代码如下：

```java
package com.bezkoder.springjwt.controllers;
import org.springframework.security.access.prepost.PreAuthorize;
import org.springframework.web.bind.annotation.CrossOrigin;
import org.springframework.web.bind.annotation.GetMapping;
import org.springframework.web.bind.annotation.RequestMapping;
import org.springframework.web.bind.annotation.RestController;
@CrossOrigin(origins = "*", maxAge = 3600)
@RestController
```

```
@RequestMapping("/api/test")
public class TestController {
    @GetMapping("/all")
    public String allAccess() {
        return "Public Content.";
    }
    @GetMapping("/user")
    @PreAuthorize("hasRole('USER') or hasRole('MODERATOR') or hasRole('ADMIN')")
    public String userAccess() {
        return "User Content.";
    }
    @GetMapping("/mod")
    @PreAuthorize("hasRole('MODERATOR')")
    public String moderatorAccess() {
        return "Moderator Board.";
    }
    @GetMapping("/admin")
    @PreAuthorize("hasRole('ADMIN')")
    public String adminAccess() {
        return "Admin Board.";
    }
}
```

使用以下命令运行 Spring Boot 应用程序。

```
>mvn spring-boot:run
```

在 models 包中定义的表将在数据库中自动生成。在将任何角色分配给 User 前，还需要向角色表中添加一些行。运行以下 SQL 插入语句：

```
INSERT INTO roles(name) VALUES('ROLE_USER');
INSERT INTO roles(name) VALUES('ROLE_MODERATOR');
INSERT INTO roles(name) VALUES('ROLE_ADMIN');
```

为了实现 Web 前端，用 React 框架实现表单验证、检查注册用户名/电子邮件重复项、具有 3 个角色（Admin、Moderator 和 User）的测试授权。接下来将构建一个 React 应用程序。

- 该程序包含"登录/注销"页面和"注册"页面。
- 表单数据在发送到后端前，将由前端进行验证。
- 导航栏会根据用户的角色（管理员、主持人、用户）自动更改其项目。

执行如下命令来建立 React.js 项目。

```
>npx create-react-app react-jwt-auth
```

执行如下命令来添加 React 路由器。

```
>npm install react-router-dom
```

打开 src/index.js 并通过 BrowserRouter 对象包装 App 组件。

```
import React from "react";
import ReactDOM from "react-dom";
import { BrowserRouter } from "react-router-dom";
import App from "./App";
import * as serviceWorker from "./serviceWorker";
```

```
ReactDOM.render(
  <BrowserRouter>
    <App />
  </BrowserRouter>,
  document.getElementById("root")
);
serviceWorker.unregister();
```

执行如下命令来导入 Bootstrap。

```
> npm install bootstrap
```

打开 src/App.js 并按照以下内容修改其中的代码。

```
import React, { Component } from "react";
import "bootstrap/dist/css/bootstrap.min.css";
class App extends Component {
  render() {
    … //省略代码
  }
}
export default App;
```

在 src/services 文件夹中创建身份验证服务和数据服务。开发这些服务前，我们需要使用以下命令安装 Axios。

```
>npm install axios
```

认证服务将 Axios 用于 HTTP 请求，将本地存储用于用户信息和 JWT。它提供了以下重要方法。

- login()：包含 post {username, password} 并且保存 JWT 到本地存储。
- logout()：从本地存储中删除 JWT。
- register()：包含 post {username, email, password}。
- getCurrentUser()：获取存储的用户信息（包括 JWT）。

认证服务实现代码如下：

```
import axios from "axios";
const API_URL = "http://localhost:8080/api/auth/";
class AuthService {
  login(username, password) {
    return axios
      .post(API_URL + "signin", {
        username,
        password
      })
      .then(response => {
        if (response.data.accessToken) {
          localStorage.setItem("user", JSON.stringify(response.data));
        }
        return response.data;
      });
  }

  logout() {
```

```
    localStorage.removeItem("user");
  }

  register(username, email, password) {
    return axios.post(API_URL + "signup", {
      username,
      email,
      password
    });
  }

  getCurrentUser() {
    return JSON.parse(localStorage.getItem('user'));;
  }
}
export default new AuthService();
```

还提供了从服务器检索数据的方法。在访问受保护资源的情况下，HTTP 请求需要 Authorization 标头。在 auth-header.js 中创建一个名为 authHeader() 的辅助函数，代码如下：

```
export default function authHeader() {
    const user = JSON.parse(localStorage.getItem('user'));
    if (user && user.accessToken) {
        return { Authorization: 'Bearer ' + user.accessToken };
    } else {
        return { };
    }
}
```

上面的代码会检查本地存储中的用户项。如果存在具有 accessToken（JWT）的登录用户，则返回 HTTP Authorization 标头；否则，返回一个空对象。

在 user.service.js 中定义用于访问数据的服务，代码如下：

```
import axios from 'axios';
import authHeader from './auth-header';
const API_URL = 'http://localhost:8080/api/test/';
class UserService {
    getPublicContent() {
        return axios.get(API_URL + 'all');
    }
    getUserBoard() {
        return axios.get(API_URL + 'user', { headers: authHeader() });
    }
    getModeratorBoard() {
        return axios.get(API_URL + 'mod', { headers: authHeader() });
    }
    getAdminBoard() {
        return axios.get(API_URL + 'admin', { headers: authHeader() });
    }
}
export default new UserService();
```

从以上代码中可以看到，当请求授权资源时，在函数 authHeader() 的帮助下添加了 HTTP

标头。

下面在 src 文件夹中创建一个名为 components 的新文件夹,并添加几个文件。需要一个用于表单验证的库,因此要向项目中添加 react-validation 库。执行命令如下:

```
>npm install react-validation validator
```

若要在此示例中使用 react-validation,需要导入以下项目。

```
import Form from "react-validation/build/form";
import Input from "react-validation/build/input";
import CheckButton from "react-validation/build/button";
import { isEmail } from "validator";
```

还需要使用验证程序中的函数 isEmail() 来验证电子邮件。以下就是将它们放入带有 validations 属性的 render() 方法中的代码。

```
const required = value => {
    if (!value) {
        return (
            <div className="alert alert-danger" role="alert">
                This field is required!
            </div>
        );
    }
};

const email = value => {
    if (!isEmail(value)) {
        return (
            <div className="alert alert-danger" role="alert">
                This is not a valid email.
            </div>
        );
    }
};

render() {
    return (
        ...
        <Form
        onSubmit={this.handleLogin}
        ref={c => {this.form = c;}}
        >
        ...
        <Input
        type="text"
        className="form-control"
        ...
        validations={[required, email]}
        />

        <CheckButton
        style={{ display: "none" }}
```

```
        ref={c => {this.checkBtn = c;}}
      />
    </Form>
    ...
  );
}
```

调用表单的 validateAll()方法来检查验证功能。然后，CheckButton 将帮助验证表单验证是否成功，因此该按钮不必显示在表单上，代码如下：

```
this.form.validateAll();
if (this.checkBtn.context._errors.length === 0) {
  // do something when no error
}
```

登录页面中有一个包含用户名和密码的表单。我们将它们当作必填字段来验证。

如果验证通过，将调用 AuthService.login()方法，然后将用户定向到"个人资料"页面：this.props.history.push("/profile");，或反馈带有响应错误的消息。

login.component.js 实现代码如下：

```
import React, { Component } from "react";
import Form from "react-validation/build/form";
import Input from "react-validation/build/input";
import CheckButton from "react-validation/build/button";
import AuthService from "../services/auth.service";
const required = value => {
  if (!value) {
    return (
      <div className="alert alert-danger" role="alert">
        This field is required!
      </div>
    );
  }
};

export default class Login extends Component {
  constructor(props) {
    super(props);
    this.handleLogin = this.handleLogin.bind(this);
    this.onChangeUsername = this.onChangeUsername.bind(this);
    this.onChangePassword = this.onChangePassword.bind(this);
    this.state = {
      username: "",
      password: "",
      loading: false,
      message: ""
    };
  }

  onChangeUsername(e) {
    this.setState({
      username: e.target.value
    });
```

```js
    }

    onChangePassword(e) {
      this.setState({
        password: e.target.value
      });
    }

    handleLogin(e) {
      e.preventDefault();
      this.setState({
        message: "",
        loading: true
      });
      this.form.validateAll();

      if (this.checkBtn.context._errors.length === 0) {
        AuthService.login(this.state.username, this.state.password).then(
          () => {
            this.props.history.push("/profile");
            window.location.reload();
          },
          error => {
            const resMessage = (error.response && error.response.data && error.response.data.message) || error.message || error.toString();

            this.setState({
              loading: false,
              message: resMessage
            });
          }
        );
      } else {
        this.setState({
          loading: false
        });
      }
    }

    render() {
      return (
        <div className="col-md-12">
          <div className="card card-container">
            <img
              src="//ssl.gstatic.com/accounts/ui/avatar_2x.png"
              alt="profile-img"
              className="profile-img-card"
            />

            <Form
              onSubmit={this.handleLogin}
              ref={c => {
```

```jsx
          this.form = c;
        }}
      >
        <div className="form-group">
          <label htmlFor="username">Username</label>
          <Input
            type="text"
            className="form-control"
            name="username"
            value={this.state.username}
            onChange={this.onChangeUsername}
            validations={[required]}
          />
        </div>

        <div className="form-group">
          <label htmlFor="password">Password</label>
          <Input
            type="password"
            className="form-control"
            name="password"
            value={this.state.password}
            onChange={this.onChangePassword}
            validations={[required]}
          />
        </div>

        <div className="form-group">
          <button
            className="btn btn-primary btn-block"
            disabled={this.state.loading}
          >
            {this.state.loading && (
              <span className="spinner-border spinner-border-sm"></span>
            )}
            <span>Login</span>
          </button>
        </div>

        {this.state.message && (
          <div className="form-group">
            <div className="alert alert-danger" role="alert">
              {this.state.message}
            </div>
          </div>
        )}
        <CheckButton
          style={{ display: "none" }}
          ref={c => {
            this.checkBtn = c;
          }}
        />
```

```
        </Form>
      </div>
    </div>
  );
 }
}
```

注册页面外观类似于登录页面。对于表单验证，需要以下更多详细信息。

- username：必填，取值范围介于 3~20 个字符。
- email：必填，表示电子邮件格式。
- password：必填，表示介于 6~40 个字符。

调用 AuthService.register()方法并显示响应消息（成功或错误）。register.component.js 实现代码如下：

```
import React, { Component } from "react";
import Form from "react-validation/build/form";
import Input from "react-validation/build/input";
import CheckButton from "react-validation/build/button";
import { isEmail } from "validator";
import AuthService from "../services/auth.service";
const required = value => {
  if (!value) {
    return (
      <div className="alert alert-danger" role="alert">
        This field is required!
      </div>
    );
  }
};

const email = value => {
  if (!isEmail(value)) {
    return (
      <div className="alert alert-danger" role="alert">
        This is not a valid email.
      </div>
    );
  }
};

const vusername = value => {
  if (value.length < 3 || value.length > 20) {
    return (
      <div className="alert alert-danger" role="alert">
        The username must be between 3 and 20 characters.
      </div>
    );
  }
};

const vpassword = value => {
```

```jsx
      if (value.length < 6 || value.length > 40) {
        return (
          <div className="alert alert-danger" role="alert">
            The password must be between 6 and 40 characters.
          </div>
        );
      }
    };

export default class Register extends Component {
  constructor(props) {
    super(props);
    this.handleRegister = this.handleRegister.bind(this);
    this.onChangeUsername = this.onChangeUsername.bind(this);
    this.onChangeEmail = this.onChangeEmail.bind(this);
    this.onChangePassword = this.onChangePassword.bind(this);
    this.state = {
      username: "",
      email: "",
      password: "",
      successful: false,
      message: ""
    };
  }
  onChangeUsername(e) {
    this.setState({
      username: e.target.value
    });
  }
  onChangeEmail(e) {
    this.setState({
      email: e.target.value
    });
  }
  onChangePassword(e) {
    this.setState({
      password: e.target.value
    });
  }
  handleRegister(e) {
    e.preventDefault();
    this.setState({
      message: "",
      successful: false
    });
    this.form.validateAll();

    if (this.checkBtn.context._errors.length === 0) {
      AuthService.register(
        this.state.username,
        this.state.email,
        this.state.password
```

```jsx
      ).then(
        response => {
          this.setState({
            message: response.data.message,
            successful: true
          });
        },
        error => {
          const resMessage = (error.response && error.response.data && error.response.data.message) || error.message || error.toString();
          this.setState({
            successful: false,
            message: resMessage
          });
        }
      );
    }
  render() {
    return (
      <div className="col-md-12">
        <div className="card card-container">
          <img
            src="//ssl.gstatic.com/accounts/ui/avatar_2x.png"
            alt="profile-img"
            className="profile-img-card"
          />

          <Form
            onSubmit={this.handleRegister}
            ref={c => {
              this.form = c;
            }}
          >
            {!this.state.successful && (
              <div>
                <div className="form-group">
                  <label htmlFor="username">Username</label>
                  <Input
                    type="text"
                    className="form-control"
                    name="username"
                    value={this.state.username}
                    onChange={this.onChangeUsername}
                    validations={[required, vusername]}
                  />
                </div>

                <div className="form-group">
                  <label htmlFor="email">Email</label>
                  <Input
                    type="text"
```

```jsx
              className="form-control"
              name="email"
              value={this.state.email}
              onChange={this.onChangeEmail}
              validations={[required, email]}
            />
          </div>

          <div className="form-group">
            <label htmlFor="password">Password</label>
            <Input
              type="password"
              className="form-control"
              name="password"
              value={this.state.password}
              onChange={this.onChangePassword}
              validations={[required, vpassword]}
            />
          </div>

          <div className="form-group">
            <button className="btn btn-primary btn-block">Sign Up</button>
          </div>
        </div>
      )}

      {this.state.message && (
        <div className="form-group">
          <div
            className={
              this.state.successful
                ? "alert alert-success"
                : "alert alert-danger"
            }
            role="alert"
          >
            {this.state.message}
          </div>
        </div>
      )}
      <CheckButton
        style={{ display: "none" }}
        ref={c => {
          this.checkBtn = c;
        }}
      />
    </Form>
   </div>
  </div>
 );
 }
}
```

个人资料页面通过调用 AuthService.getCurrentUser()方法从本地存储获取当前用户，并显示用户信息（带有令牌）。

profile.component.js 实现代码如下：

```js
import React, { Component } from "react";
import AuthService from "../services/auth.service";
export default class Profile extends Component {
  constructor(props) {
    super(props);
    this.state = {
      currentUser: AuthService.getCurrentUser()
    };
  }

  render() {
    const { currentUser } = this.state;
    return (
      <div className="container">
        <header className="jumbotron">
          <h3>
            <strong>{currentUser.username}</strong> Profile
          </h3>
        </header>
        <p>
          <strong>Token:</strong>{" "}
          {currentUser.accessToken.substring(0, 20)} …{" "}
          {currentUser.accessToken.substr(currentUser.accessToken.length - 20)}
        </p>
        <p>
          <strong>Id:</strong>{" "}
          {currentUser.id}
        </p>
        <p>
          <strong>Email:</strong>{" "}
          {currentUser.email}
        </p>
        <strong>Authorities:</strong>
        <ul>
          {currentUser.roles &&
            currentUser.roles.map((role, index) => <li key={index}>{role}</li>)}
        </ul>
      </div>
    );
  }
}
```

创建 React 组件以访问资源，这些组件将使用 UserService 从 API 请求数据。

主页是显示公共内容的公共页面，无须登录即可查看此页面。home.component.js 实现代码如下：

```js
import React, { Component } from "react";
import UserService from "../services/user.service";
```

```
export default class Home extends Component {
  constructor(props) {
    super(props);
    this.state = {
      content: ""
    };
  }

  componentDidMount() {
    UserService.getPublicContent().then(
      response => {
        this.setState({
          content: response.data
        });
      },
      error => {
        this.setState({
          content:
            (error.response && error.response.data) || error.message || error.toString()
        });
      }
    );
  }

  render() {
    return (
      <div className="container">
        <header className="jumbotron">
          <h3>{this.state.content}</h3>
        </header>
      </div>
    );
  }
}
```

有以下 3 个基于角色的页面将用于访问受保护的数据。

- BoardUser 页面：调用 UserService.getUserBoard()。
- BoardModerator 页面：调用 UserService.getModeratorBoard()。
- BoardAdmin 页面：调用 UserService.getAdminBoard()。

这里只展示用户页面的代码，其他页面与此页面相似。board-user.component.js 实现代码如下：

```
import React, { Component } from "react";
import UserService from "../services/user.service";
export default class BoardUser extends Component {
  constructor(props) {
    super(props);
    this.state = {
      content: ""
    };
  }
```

```js
  componentDidMount() {
    UserService.getUserBoard().then(
      response => {
        this.setState({
          content: response.data
        });
      },
      error => {
        this.setState({
          content:
            (error.response && error.response.data && error.response.data.message) ||
            error.message || error.toString()
        });
      }
    );
  }

  render() {
    return (
      <div className="container">
        <header className="jumbotron">
          <h3>{this.state.content}</h3>
        </header>
      </div>
    );
  }
}
```

在 App 组件中添加一个导航栏（这是应用程序的根容器）。导航栏会根据登录状态和当前用户的角色动态变化，其主要包括以下界面等项目。

- 主页：总是存在。
- 登录和注册：如果用户尚未登录，会显示登录和注册。
- 用户：AuthService.getCurrentUser()返回一个值。
- 主持人面板：包括 ROLE_MODERATOR 的角色。
- 管理员面板：包括 ROLE_ADMIN 的角色。

src/App.js 实现代码如下：

```js
import React, { Component } from "react";
import { BrowserRouter as Router, Switch, Route, Link } from "react-router-dom";
import "bootstrap/dist/css/bootstrap.min.css";
import "./App.css";
import AuthService from "./services/auth.service";
import Login from "./components/login.component";
import Register from "./components/register.component";
import Home from "./components/home.component";
import Profile from "./components/profile.component";
import BoardUser from "./components/board-user.component";
import BoardModerator from "./components/board-moderator.component";
import BoardAdmin from "./components/board-admin.component";
```

```
class App extends Component {
  constructor(props) {
    super(props);
    this.logOut = this.logOut.bind(this);
    this.state = {
      showModeratorBoard: false,
      showAdminBoard: false,
      currentUser: undefined
    };
  }

  componentDidMount() {
    const user = AuthService.getCurrentUser();
    if (user) {
      this.setState({
        currentUser: user,
        showModeratorBoard: user.roles.includes("ROLE_MODERATOR"),
        showAdminBoard: user.roles.includes("ROLE_ADMIN")
      });
    }
  }

  logOut() {
    AuthService.logout();
  }

  render() {
    const { currentUser, showModeratorBoard, showAdminBoard } = this.state;
    return (
      <Router>
        <div>
          <nav className="navbar navbar-expand navbar-dark bg-dark">
            <Link to={"/"} className="navbar-brand">
              bezKoder
            </Link>
            <div className="navbar-nav mr-auto">
              <li className="nav-item">
                <Link to={"/home"} className="nav-link">
                  Home
                </Link>
              </li>

              {showModeratorBoard && (
                <li className="nav-item">
                  <Link to={"/mod"} className="nav-link">
                    Moderator Board
                  </Link>
                </li>
              )}

              {showAdminBoard && (
```

```jsx
              <li className="nav-item">
                <Link to={"/admin"} className="nav-link">
                  Admin Board
                </Link>
              </li>
          )}

          {currentUser && (
            <li className="nav-item">
              <Link to={"/user"} className="nav-link">
                User
              </Link>
            </li>
          )}
        </div>

        {currentUser ? (
          <div className="navbar-nav ml-auto">
            <li className="nav-item">
              <Link to={"/profile"} className="nav-link">
                {currentUser.username}
              </Link>
            </li>
            <li className="nav-item">
              <a href="/login" className="nav-link" onClick={this.logOut}>
                LogOut
              </a>
            </li>
          </div>
        ) : (
          <div className="navbar-nav ml-auto">
            <li className="nav-item">
              <Link to={"/login"} className="nav-link">
                Login
              </Link>
            </li>

            <li className="nav-item">
              <Link to={"/register"} className="nav-link">
                Sign Up
              </Link>
            </li>
          </div>
        )}
      </nav>

      <div className="container mt-3">
        <Switch>
          <Route exact path={["/", "/home"]} component={Home} />
          <Route exact path="/login" component={Login} />
          <Route exact path="/register" component={Register} />
          <Route exact path="/profile" component={Profile} />
```

```
              <Route path="/user" component={BoardUser} />
              <Route path="/mod" component={BoardModerator} />
              <Route path="/admin" component={BoardAdmin} />
          </Switch>
        </div>
      </div>
    </Router>
  );
 }
}

export default App;
```

为 React 组件添加 CSS 样式。打开 src/App.css 文件，并编写一些 CSS 代码，代码如下：

```
label {
  display: block;
  margin-top: 10px;
}

.card-container.card {
  max-width: 350px !important;
  padding: 40px 40px;
}

.card {
  background-color: #f7f7f7;
  padding: 20px 25px 30px;
  margin: 0 auto 25px;
  margin-top: 50px;
  -moz-border-radius: 2px;
  -webkit-border-radius: 2px;
  border-radius: 2px;
  -moz-box-shadow: 0px 2px 2px rgba(0, 0, 0, 0.3);
  -webkit-box-shadow: 0px 2px 2px rgba(0, 0, 0, 0.3);
  box-shadow: 0px 2px 2px rgba(0, 0, 0, 0.3);
}

.profile-img-card {
  width: 96px;
  height: 96px;
  margin: 0 auto 10px;
  display: block;
  -moz-border-radius: 50%;
  -webkit-border-radius: 50%;
  border-radius: 50%;
}
```

由于大多数 HTTP Server 使用 CORS 配置来接受受限于某些站点或端口的资源共享，因此还需要为应用配置端口。

在项目文件夹中，创建具有以下内容的 .env 文件。

```
PORT=8081
```

在这里，我们将应用程序设置为在端口 8081 上运行。

4.4 OAuth 2 授权框架

OAuth 2 是用于访问委派的开放标准，通常用作 Internet 用户授予网站或应用程序访问其在其他网站上所存储信息的一种权限方式，而无须提供密码。Keycloak 是面向现代应用程序与服务的开源身份和访问管理解决方案。资源服务器是 API 服务器的 OAuth 2 术语。在应用程序获取访问令牌后，资源服务器将处理经过身份验证的请求。下面先介绍使用 Spring Boot 搭建资源服务器。

4.4.1 OAuth 2 资源服务器和 Keycloak 服务器

本小节介绍如何实现一个非常简单的 OAuth 2 资源服务器，该服务器将验证它先前从 Keycloak 授权服务器获取的 JWT 令牌。

首先需要一个安装好的 Keycloak 服务器（可以从官方网站下载 Keycloak 服务器）。从 Keycloak 发行版的 bin/目录中，我们可以通过运行 standalone.sh 脚本（macOS 和 Linux 操作系统）或 standalone.bat 文件（Windows 操作系统）来启动 Keycloak 服务器。默认情况下，Keycloak 服务器将在主机 8080 端口上可访问，但是我们可以通过设置 jboss.socket.binding.port-offset 属性来自定义端口。在 Keycloak 管理控制台可以创建新领域、创建角色和定义默认角色、添加新用户及为其分配密码和角色。

为了使 Spring Boot 项目能够用作资源服务器，并且能够与 Keycloak 服务器进行通信以验证 JWT，需要向其添加一个非常重要的依赖项 spring-boot-starter-oauth2-resource-server。打开 pom.xml 文件，并添加以下依赖项。

```
<dependency>
    <groupId>org.springframework.boot</groupId>
    <artifactId>spring-boot-starter-oauth2-resource-server</artifactId>
</dependency>
```

上面的依赖关系已经包含 Spring Security 库，因此无须向此 pom.xml 文件添加任何其他 Spring Security 依赖关系。

需要添加以使 Spring Boot 项目作为 RESTful Web Service 应用程序工作的唯一附加依赖项是 spring-boot-starter-web。

```
<dependency>
    <groupId>org.springframework.boot</groupId>
    <artifactId>spring-boot-starter-web</artifactId>
</dependency>
```

以下是资源服务器项目的完整 pom.xml 文件。

```
<?xml version="1.0" encoding="UTF-8"?>
<project xmlns="http://maven.apache.org/POM/4.0.0" xmlns:xsi="http://www.w3.org/
2001/XMLSchema-instance"
```

```xml
        xsi:schemaLocation="http://maven.apache.org/POM/4.0.0 https://maven.apache.
org/xsd/maven-4.0.0.xsd">
    <modelVersion>4.0.0</modelVersion>
    <parent>
        <groupId>org.springframework.boot</groupId>
        <artifactId>spring-boot-starter-parent</artifactId>
        <version>2.3.0.RELEASE</version>
        <relativePath/> <!-- lookup parent from repository -->
    </parent>
    <groupId>com.appsdeveloperblog.keycloak</groupId>
    <artifactId>KeycloakResourceServer2</artifactId>
    <version>0.0.1-SNAPSHOT</version>
    <name>KeycloakResourceServer2</name>
    <description>Demo project for Spring Boot</description>
    <properties>
        <java.version>1.8</java.version>
    </properties>
    <dependencies>
        <dependency>
            <groupId>org.springframework.boot</groupId>
            <artifactId>spring-boot-starter-oauth2-resource-server</artifactId>
        </dependency>
        <dependency>
            <groupId>org.springframework.boot</groupId>
            <artifactId>spring-boot-starter-web</artifactId>
        </dependency>
        <dependency>
            <groupId>org.springframework.boot</groupId>
            <artifactId>spring-boot-starter-test</artifactId>
            <scope>test</scope>
            <exclusions>
                <exclusion>
                    <groupId>org.junit.vintage</groupId>
                    <artifactId>junit-vintage-engine</artifactId>
                </exclusion>
            </exclusions>
        </dependency>
    </dependencies>
    <build>
        <plugins>
            <plugin>
                <groupId>org.springframework.boot</groupId>
                <artifactId>spring-boot-maven-plugin</artifactId>
            </plugin>
        </plugins>
    </build>
</project>
```

为了使 Spring Boot 应用程序能够作为 RESTful Web Service 应用程序工作，下面将创建一个简单的@RestController。在此@RestController 类的内部创建一个非常简单的 Web 服务端点，该端点将返回纯文本作为响应。

```
import org.springframework.web.bind.annotation.GetMapping;
import org.springframework.web.bind.annotation.RequestMapping;
import org.springframework.web.bind.annotation.RestController;
@RestController
@RequestMapping("/users")
public class UsersController {
    @GetMapping("/status/check")
    public String status() {
        return "working";
    }
}
```

如果运行此 Spring Boot 应用程序，并尝试通过 http://localhost:8080/users/status/check 访问此 Web 服务端点，则会接收到 401 未经授权的状态代码。这是因为在项目中添加了 spring-boot-starter-oauth2-resource-server 依赖项。此时，Web 服务端点已受到保护，要访问它，就需要提供一个有效的访问令牌。例如，为了使客户端应用程序能够与资源服务器进行通信，它们将需要包含由授权服务器生成的有效 JWT 令牌。在这里，使用 Keycloak 服务器作为授权服务器，而将要使用的 JWT 令牌是由 Keycloak 服务器发行的。

为了让资源服务器与 JWT 的发布者一起验证 JWT，需要向项目的 application.properties 文件中添加一个属性，代码如下：

```
spring.security.oauth2.resourceserver.jwt.issuer-uri = http://localhost:8080/auth/realms/appsdeveloperblog
```

这里的 issuer-uri 值指向在 Keycloak 服务器中创建的 appsdeveloperblog 领域。

配置资源服务器和授权服务器验证 JWT 令牌的另一种方法是使用以下属性。

```
spring.security.oauth2.resourceserver.jwt.jwk-set-uri = http://localhost:8080/auth/realms/appsdeveloperblog/protocol/openid-connect/certs
```

上面的 jwk-set-uri 值指向公钥端点。如果在浏览器中打开此 Web 服务端点，将看到一个 JSON，其中包含将用于验证 JWT 的信息。

可以使用 spring.security.oauth2.resourceserver.jwt.issuer-uri 或 spring.security.oauth2.resourceserver.jwt.jwk-set-uri，甚至可以同时使用它们。如果两个属性都使用且其中一个的值不正确，则 JWT 的验证将失败。

```
spring.security.oauth2.resourceserver.jwt.issuer-uri = http://localhost:8080/auth/realms/appsdeveloperblog
spring.security.oauth2.resourceserver.jwt.jwk-set-uri = http://localhost:8080/auth/realms/appsdeveloperblog/protocol/openid-connect/certs
```

要访问资源服务器中具有的任何 Web 服务端点，则需要在 HTTP 请求的授权标头中包含有效的 JWT。以下是用 curl 命令来访问已创建的 Web 服务端点的示例。

```
curl --location --request GET 'http://localhost:8082/users/status/check' \
--header 'Authorization: Bearer eyJhbGciOiJSUzI1NiIsInR5cCIgOiAiSldUIiwia2lkIiA6 ICItNUlsX2I0cUktdWFvaEI3d244UHY3WEM2UEktU3BNbmZCRnlJZUx6QTJNIn0.eyJleHAiOjE1OTI0MTc4 TcsImlhdCI6MT…'
```

4.4.2　Spring Security 和 Keycloak 保护 Spring Boot 应用程序

本小节将引入 Spring Security，并了解如何将其与 Keycloak 无缝集成来保护 Spring Boot 应用程序。将要构建的应用程序是针对虚拟公共图书馆的。其有以下两类用户：
- 成员，可以浏览图书馆中可用的书籍；
- 图书管理员，还可以管理书籍。

对于每组用户，在 Keycloak 中定义了一个相应的用户角色 Member 和 Librarian。此外，在系统中注册了两个用户：Sheldon 是成员，Irma 既是成员又是图书馆员。该应用程序将利用以下 3 个主要的库来设置 Spring。
- spring-boot-starter-web：一个用于使用 Spring MVC 构建 Web 应用程序的启动器。
- spring-boot-starter-thymeleaf：一个将 Thymeleaf 视图用于 Spring MVC 的启动器。
- spring-boot-starter-security：一个使用 Spring Security 的启动程序。

通过客户端适配器来与 Keycloak 集成。Keycloak 客户端适配器是一个库，可以通过 Keycloak 轻松保护应用程序和服务的安全。

此项目还需要：
- Spring Boot Adapter 利用 Spring Boot 的自动配置功能；
- Spring Security Adapter 来使用 Keycloak 作为 Spring Security 的身份验证提供程序库。

keycloak-spring-boot-starter 库同时包含了这两个库，因此不需要其他任何库。

可以在 gradle.build 文件中为项目定义所有这些依赖项（如果使用 Maven,则可以在 pom.xml 中定义），代码如下：

```
buildscript {
    ext {
        springBootVersion = '2.1.9.RELEASE'
    }
    repositories {
        mavenCentral()
    }
    dependencies {
classpath("org.springframework.boot:spring-boot-gradle-plugin:${springBootVersion}")
    }
}

apply plugin: 'java'
apply plugin: 'org.springframework.boot'
apply plugin: 'io.spring.dependency-management'
group = 'com.thomasvitale'
version = '0.0.1-SNAPSHOT'
sourceCompatibility = '1.8'
repositories {
    mavenCentral()
}
ext {
    set('keycloakVersion', '7.0.1')
}
```

```
dependencies {
   // Spring
   implementation 'org.springframework.boot:spring-boot-starter-web'
   implementation 'org.springframework.boot:spring-boot-starter-thymeleaf'
   implementation 'org.springframework.boot:spring-boot-starter-security'
   // Keycloak
   implementation 'org.keycloak:keycloak-spring-boot-starter'
   // Test
   testImplementation 'org.springframework.boot:spring-boot-starter-test'
   testImplementation 'org.springframework.boot:spring-security-test'
   testImplementation 'org.keycloak:keycloak-test-helper'
}

dependencyManagement {
   imports {
      mavenBom "org.keycloak.bom:keycloak-adapter-bom:${keycloakVersion}"
   }
}
```

至此，已经建立了一个 Spring Boot 应用程序，并准备好了所要利用的 Spring Security 和 Keycloak。

接下来，应该为 Spring Boot 和 Keycloak 之间的集成提供一些配置。Keycloak 的默认配置文件是 keycloak.json，但是由于有了 Keycloak Spring Boot Adapter，可以使用本机的 application.properties 文件（或 application.yml）。例如：

```
keycloak.realm=public-library
keycloak.resource=spring-boot-app
keycloak.auth-server-url=http://localhost:8180/auth
keycloak.ssl-required=external
keycloak.public-client=true
keycloak.principal-attribute=preferred_username
```

下面一一了解这些属性。

- keycloak.realm：必填，领域的名称。
- keycloak.resource：必填，应用程序的客户端 ID。
- keycloak.auth-server-url：必填，Keycloak 服务器的基本 URL。
- keycloak.ssl-required：可选的，确定是否必须通过 HTTPS 与 Keycloak 服务器进行通信。在这里，将其设置为 external，这意味着仅对于外部请求才需要该默认值。在生产中，应该将其设置为 all。
- keycloak.public-client：可选的，防止应用程序将凭据发送到 Keycloak 服务器（默认值为 false）。每当使用公共客户端而不是机密客户端时，都希望将其设置为 true。
- keycloak.principal-attribute：可选的，用于填充 UserPrincipal 名称的属性。

在 @Configuration 类中定义 KeycloakSpringBootConfigResolver Bean，代码如下：

```
@Configuration
public class KeycloakConfig {

    @Bean
```

```
    public KeycloakSpringBootConfigResolver keycloakConfigResolver() {
        return new KeycloakSpringBootConfigResolver();
    }
}
```

想要使用 Spring Security 和 Keycloak，在访问某些受保护的端点之前确保对应用程序的用户进行身份验证和授权。

接着，在 Spring MVC Controller 类中定义以下 3 个端点。
- /index：将可以自由访问。
- /books：仅具有成员角色的用户可以访问，例如他们可以浏览图书馆中可用的书。
- /manager：仅具有图书管理员角色的用户可以访问，例如他们可以管理图书。

稍后，将配置最后两个端点，以要求用户都经过身份验证并具有适当的角色。添加/logout 端点以方便注销。

其实现代码如下：

```
@Controller
public class LibraryController {
    private final HttpServletRequest request;
    private final BookRepository bookRepository;
    @Autowired
    public LibraryController(HttpServletRequest request, BookRepository bookRepository) {
        this.request = request;
        this.bookRepository = bookRepository;
    }
    @GetMapping(value = "/")
    public String getHome() {
        return "index";
    }
    @GetMapping(value = "/books")
    public String getBooks(Model model) {
        configCommonAttributes(model);
        model.addAttribute("books", bookRepository.readAll());
        return "books";
    }
    @GetMapping(value = "/manager")
    public String getManager(Model model) {
        configCommonAttributes(model);
        model.addAttribute("books", bookRepository.readAll());
        return "manager";
    }

    @GetMapping(value = "/logout")
    public String logout() throws ServletException {
        request.logout();
        return "redirect:/";
    }

    private void configCommonAttributes(Model model) {
        model.addAttribute("name", getKeycloakSecurityContext()
.getIdToken().getGivenName());
```

```
        }
        private KeycloakSecurityContext getKeycloakSecurityContext() {
            return (KeycloakSecurityContext) request.getAttribute(KeycloakSecurityContext.class.getName());
        }
    }
```

在最后一个方法中,使用 KeycloakSecurityContext 检索 IdToken,从中可以获取经过身份验证的用户名称。

这里使用 Thymeleaf 模板引擎。为每种资源都准备了一个模板,还有一个用于处理未经授权请求的唯一模板。

LibraryController 类将 Book 实体的获取委托给 BookRepository 类,代码如下:

```
@Repository
public class BookRepository {
    private static Map<String, Book> books = new ConcurrentHashMap<>();
    static {
        books.put("B01", new Book("B01", "Harry Potter and the Deathly Hallows", "J.K. Rowling"));
        books.put("B02", new Book("B02", "The Lord of the Rings", "J.R.R. Tolkien"));
        books.put("B03", new Book("B03", "War and Peace", "Leo Tolstoy"));
    }
    public List<Book> readAll() {
        List<Book> allBooks = new ArrayList<>(books.values());
        allBooks.sort(Comparator.comparing(Book::getId));
        return allBooks;
    }
}
```

Book 实体类是 POJO,其实现代码如下:

```
public class Book {
    private String id;
    private String title;
    private String author;
    public Book(String id, String title, String author) {
        this.id = id;
        this.title = title;
        this.author = author;
    }
    …// Getters and setters
}
```

如果此时尝试运行该应用程序,则会注意到可以无限制地浏览所有页面。下面改变一下相关配置等。

例如,将 Spring Security 配置为:

- 将用户身份验证阶段委托给 Keycloak;
- 利用 Keycloak 返回的 IdToken 和 AccessToken 对象为应用程序端点定义一些访问策略。

从创建 SecurityConfig 类开始逐步进行操作,代码如下:

```
@KeycloakConfiguration
public class SecurityConfig extends KeycloakWebSecurityConfigurerAdapter {
```

```
...// Configuration
}
```

SecurityConfig 类应扩展 KeycloakWebSecurityConfigurerAdapter 并使用@KeycloakConfiguration 进行注解。该注解提供了基于 Keycloak 的 Spring Security 配置。封装 Spring Security 配置类所需的两个注解@Configuration 和@EnableWebSecurity 是一种固定写法。它还包装了 Keycloak 所需的第三个注解，以正确扫描在 Keycloak Spring Security 适配器中配置的 Bean：@ComponentScan (basePackageClasses = KeycloakSecurityComponents.class)。

Keycloak 是身份提供者（IdP），需在 Spring Security 身份验证管理器中注册它，代码如下：

```
@Autowired
public void configureGlobal(AuthenticationManagerBuilder auth) {
    SimpleAuthorityMapper grantedAuthorityMapper = new SimpleAuthorityMapper();
    grantedAuthorityMapper.setPrefix("ROLE_");
    KeycloakAuthenticationProvider keycloakAuthenticationProvider = keycloakAuthenticationProvider();
    keycloakAuthenticationProvider.setGrantedAuthoritiesMapper(grantedAuthorityMapper);
    auth.authenticationProvider(keycloakAuthenticationProvider);
}
```

在这里，正在向身份验证管理器注册 KeycloakAuthenticationProvider。这样，Keycloak 将负责提供身份验证服务。

Spring Security 有一个约定，即以 ROLE_ADMIN 等格式处理安全角色（其中 ADMIN 是实际的安全角色名称）。在 Keycloak 中，笔者更喜欢定义首字母大写的角色名称。为了解决我们定义用户角色的方式与 Spring Security 使用的约定之间的不匹配问题，可以指定其他配置。因此，为 KeycloakAuthenticationProvider 设置了一个 SimpleAuthorityMapper，以便为在 Keycloak 中注册的所有角色添加前缀。这将有助于 Spring Security 以最佳方式处理这些角色。

在 Keycloak 中定义了两个角色：Member 和 Librarian。在 Spring Security 中映射后，它们将变为 ROLE_Member 和 ROLE_Librarian。我们甚至可以通过调用方法 grantAuthorityMapper.setConvertToUpperCase(true);将完整的角色名称变为大写，但是对于此演示应用程序，我们不这样做。

用 Keycloak 的术语来说，正在构建的应用程序是带有用户交互的公共应用程序。在这种情况下，建议的会话身份验证策略是 RegisterSessionAuthenticationStrategy，该策略在身份验证成功后注册用户会话，代码如下：

```
@Bean
@Override
protected SessionAuthenticationStrategy sessionAuthenticationStrategy() {
    return new RegisterSessionAuthenticationStrategy(new SessionRegistryImpl());
}
```

在保护服务到服务的应用程序时，将使用 NullAuthenticatedSessionStrategy。

到目前为止，已经完成了所有必要的配置，以使 Spring Security 与 Keycloak 无缝协作。我们的最后一步是为应用程序端点定义一些安全性约束。

```
@Override
protected void configure(HttpSecurity http) throws Exception {
    super.configure(http);
    http
```

```
            .authorizeRequests()
                .antMatchers("/books").hasAnyRole("Member", "Librarian")
                .antMatchers("/manager").hasRole("Librarian")
                .anyRequest().permitAll();
    }
```

由以上代码可知，应用程序端点的访问策略如下。
- /books：用户必须经过身份验证，并且至少具有成员和图书管理员中的一种角色。
- /manager：用户必须经过身份验证并具有图书管理员角色。
- 任何其他端点都可以自由访问，没有角色约束，不需要身份验证。

注意：从 Spring 2.1.0 开始，spring.main.allow-bean-definition-overriding 属性默认值设置为 false，与以前的 Spring 版本不同。这意味着不再允许重写已定义的 Bean。

SecurityConfig 类扩展了 KeycloakWebSecurityConfigurerAdapter，它定义了 HttpSessionManager Bean。但是，此 Bean 已在 Keycloak 适配器库中的其他位置定义。 因此，它在 Spring 2.1.0 以上版本中会触发错误。

```
***************************
APPLICATION FAILED TO START
***************************

Description:
The bean 'httpSessionManager', defined in class path resource [com/thomasvitale/
keycloak/config/SecurityConfig.class], could not be registered. A bean with that name has
already been defined in URL [··· /org/keycloak/adapters/springsecurity/management/
HttpSessionManager.class] and overriding is disabled.

Action:
Consider renaming one of the beans or enabling overriding by setting spring.
main.allow-bean-definition-overriding=true

Process finished with exit code 1
```

可以通过将 spring.main.allow-bean-definition-overriding 的值更改为 true 来解决此问题。这里将 Bean 定义调整为像以下这样有条件地加载，只有在没有其他类型的 Bean 被定义时，才可以加载。

```
@Bean
@Override
@ConditionalOnMissingBean(HttpSessionManager.class)
protected HttpSessionManager httpSessionManager() {
    return new HttpSessionManager();
}
```

至此，已经完成了安全性配置，其最终代码如下：

```
@KeycloakConfiguration
public class SecurityConfig extends KeycloakWebSecurityConfigurerAdapter {
    @Autowired
    public void configureGlobal(AuthenticationManagerBuilder auth) {
        SimpleAuthorityMapper grantedAuthorityMapper = new SimpleAuthorityMapper();
```

```
            grantedAuthorityMapper.setPrefix("ROLE_");
            KeycloakAuthenticationProvider keycloakAuthenticationProvider = keycloak
AuthenticationProvider();
            keycloakAuthenticationProvider.setGrantedAuthoritiesMapper(grantedAuthority
Mapper);
            auth.authenticationProvider(keycloakAuthenticationProvider);
    }
    @Bean
    @Override
    protected SessionAuthenticationStrategy sessionAuthenticationStrategy() {
        return new RegisterSessionAuthenticationStrategy(new SessionRegistryImpl());
    }
    @Bean
    @Override
    @ConditionalOnMissingBean(HttpSessionManager.class)
    protected HttpSessionManager httpSessionManager() {
        return new HttpSessionManager();
    }
    @Override
    protected void configure(HttpSecurity http) throws Exception {
        super.configure(http);
        http
            .authorizeRequests()
            .antMatchers("/books").hasAnyRole("Member", "Librarian")
            .antMatchers("/manager").hasRole("Librarian")
            .anyRequest().permitAll();
    }
}
```

验证我们的代码是否可以按预期工作。可能已经注意到，在设置 build.gradle 时定义了一些用于测试的依赖项。

- spring-boot-starter-test：用于通过 JUnit、Hamcrest 和 Mockito 测试 Spring Boot 应用程序的启动程序。
- spring-security-test：提供实用程序来测试 Spring Security。
- keycloak-test-helper：有助于使用 Keycloak 测试应用程序。

这里不编写自动测试的代码，而是运行应用程序并手动检查它是否正常运行。确保 Keycloak 服务器已启动并正在运行，启动 Spring Boot 应用程序。默认情况下，Spring Boot 应用程序将在 http://localhost:8080 上可用。接下来，进行相应测试。

①主页可自由访问。我们不需要通过身份验证，也不需要扮演特定角色。当导航到其他页面时，应用程序会将我们重定向到 Keycloak 进行登录。

②提供正确的用户名和密码后，Keycloak 将我们重定向回应用程序。根据所拥有的角色，我们也可以访问其他页面。

③如果以 Sheldon Cooper 等会员身份登录，则可以查看"浏览书籍"页面。即使已通过身份验证，也无权浏览到"管理库"页面。如果尝试这样做，则会被告知不允许进入页面。

④如果以 Irma Pince 等图书管理员身份登录，则可以看到"管理库"页面。

⑤注意右上角的"注销"链接，可以使用该链接进行注销。

经上述测试，该应用程序能够正常运行，并且所有安全约束均已得到实施。

4.5 本章小结

本章介绍了使用 Spring Security 和 Apache Shiro 实现安全管理，以及集成 JWT 身份验证的 Spring Boot 应用程序。

此外，Keycloak 可同时支持 OpenID Connect（OAuth 2.0 的扩展）和 SAML 2.0。保护客户端和服务安全时，需要决定的第一件事就是要使用两者中的哪一个。根据需要，还可以选择使用 OpenID Connect 保护某些安全性，并使用 SAML 保护其他安全性。为了保护客户端和服务的安全，还需要选择适配器或库。Keycloak 带有针对特定平台的适配器，但是也可以使用通用的 OpenID Connect 依赖方库和 SAML 服务提供程序库。

第 5 章

Spring Boot 整合搜索引擎

Solr 是一个由商用系统演化出来的、基于 Lucene 的开源搜索引擎项目。Elasticsearch 是一个分布式的、高可用的搜索引擎。本章介绍 Spring Boot 通过整合 Solr 或者 Elasticsearch 来提供搜索功能。

5.1 用于 Solr 的 Spring Data

Spring Data 提供了一套数据访问层（DAO）的解决方案，致力于减少数据访问层的开发量。Spring Data Solr（https://github.com/spring-projects/spring-data-solr）是 Spring Data 的子项目。

控制类 SearchSolr 可以通过 Spring Data Solr 中的工厂类 HttpSolrClientFactoryBean 得到 SolrClient 对象。配置类 SolrConfig 实现代码如下：

```
@Configuration
@EnableSolrRepositories(basePackages = {"solrapp"})
public class SolrConfig {
   @Value("${solr.url}")
   private String host;
   private volatile static SolrClient INSTANCE = null;   //用于实现单件模式
   private static final Object mutex = new Object();
   @Bean
   public SolrClient solrClient() throws Exception {
      if (INSTANCE != null) {
         return INSTANCE;
      }
      synchronized (mutex) {
         HttpSolrClientFactoryBean factory = new HttpSolrClientFactoryBean();
         factory.setUrl(host);
         factory.afterPropertiesSet();
         INSTANCE = factory.getSolrClient();
      }
      return INSTANCE;
   }
   @Bean
   public SolrTemplate solrTemplate(SolrClient solrClient) throws Exception {
      SolrTemplate solrTemplate = new SolrTemplate(solrClient);
      return solrTemplate;
   }
```

Spring Data 以一个命名为 Repository 的接口类为基础。Repository 定义代码如下：

```
public interface Repository<T, ID extends Serializable> {
}
```

Repository 是访问底层数据模型的超级接口。而对于某种具体的数据访问操作，则在其子接口中定义。例如，Spring Data Solr 项目中定义了访问 Solr 中数据的 SolrRepository 接口。而 SimpleSolrRepository 扩展了 PagingAndSortingRepository，它为分页和排序提供了内置的支持。

在 build.gradle 文件中增加以下依赖项。

```
dependencies {
    compile 'org.springframework:spring-core:4.3.25.RELEASE'
    compile 'org.springframework.data:spring-data-solr:2.0.5.RELEASE'
    compile 'org.springframework:spring-context:4.3.25.RELEASE'
    compile 'org.slf4j:slf4j-api:1.7.21'
    compile 'ch.qos.logback:logback-classic:1.1.7'
    compile 'ch.qos.logback:logback-core:1.1.7'
    compile 'org.slf4j:jcl-over-slf4j:1.7.21'
}
```

定义一个名为 Product 的文档，代码如下：

```
@SolrDocument(solrCoreName = "product")
public class Product {
    @Id
    @Indexed(name = "id", type = "string")
    private String id;

    @Indexed(name = "name", type = "string")
    private String name;
}
```

在以上代码中，@SolrDocument 注解指明 Product 类是 Solr 文档，并索引到名为 Product 的核。用@Indexed 注解的字段会在 Solr 中建立索引，并且可以搜索该字段。

接下来，需要通过扩展 Spring Data Solr 提供的存储库来创建存储库接口。使用 Product 和 String 作为实体 ID 来对存储库进行参数化，代码如下：

```
@Configuration
@EnableSolrRepositories(basePackages = "com.lietu.spring.data.solr.repository", namedQueriesLocation = "classpath:solr-named-queries.properties", multicoreSupport = true)
@ComponentScan
public class SolrConfig {
    @Bean
    public SolrClient solrClient() {
        return new HttpSolrClient("http://localhost:8983/solr");
    }
    @Bean
    public SolrTemplate solrTemplate(SolrClient client) throws Exception {
        return new SolrTemplate(client);
    }
}
```

以上代码表示正在使用@EnableSolrRepositories 来扫描包中的存储库。注意，这里指定了命名查询属性文件的位置并启用了多核支持。如果未启用多核，那么在默认情况下，Spring Data 将假定 Solr 配置用于单个核。

使用存储库索引单个文档，代码如下：

```
Product phone = new Product();        //创建实体类
phone.setId("P0001");
phone.setName("Phone");
productRepository.save(phone);        //保存实体类到存储库
```

检索并部分更新文档，代码如下：

```
Product retrievedProduct = productRepository.findOne("P0001");
retrievedProduct.setName("Smart Phone");
productRepository.save(retrievedProduct);
```

只需调用 delete 方法即可删除文档，代码如下：

```
productRepository.delete(retrievedProduct);
```

可以通过将查询放在方法的@Query 注解中来创建 Solr 搜索查询。findByCustomQuery()方法带有@Query 注解，代码如下：

```
@Query("id:*?0* OR name:*?0*")
public Page<Product> findByCustomQuery(String searchTerm, Pageable pageable);
```

使用 findByCustomQuery()方法来检索文档，代码如下：

```
Page<Product> result
  = productRepository.findByCustomQuery("Phone", new PageRequest(0, 10));
```

在以上代码中，通过调用 findByCustomQuery("Phone", new PageRequest(0, 10))，获得了产品文档的第一页。

5.2 用于 Elasticsearch 的 Spring Data

Spring Data Elasticsearch 是 Spring Data 的子项目。它实现了 Spring Data 访问 Elasticsearch 存储并提供了 Spring Data JPA（Java 持久化 API）模型的访问方式。

Repository 定义代码如下：

```
public interface Repository<T, ID extends Serializable> {
}
```

Spring Data Elasticsearch 项目中定义了访问 Elasticsearch 中数据的 ElasticsearchRepository 接口。ElasticsearchRepository 扩展了 PagingAndSortingRepository，它为分页和排序提供了内置的支持。

所有继承 Repository 接口的界面都由 Spring 管理，此接口作为标识接口，功能就是用来控制领域模型。Spring Data 可以只定义接口，只要遵循 Spring Data 的规范，就无须编写实现类代码。Spring 可以根据接口中定义的方法名实现 Repository。

要构建 RESTful API，必须了解以下 4 件事情。

- 控制器：控制器控制 HTTP 请求与应用程序逻辑之间的交互。
- 资源：指定连接到 ES 服务器的参数。
- 链接：翻页链接。
- 如何构建这些链接：在 REST 控制器中使用 spring-data-common 的 PagedResourcesAssembler，它可以在响应中生成下一页/上一页的链接。

引入包 spring-data-elasticsearch，并且去掉依赖项 elasticsearch，代码如下：

```xml
<dependency>
    <groupId>org.springframework.data</groupId>
    <artifactId>spring-data-elasticsearch</artifactId>
    <version>${spring-data-elasticsearch-version}</version>
    <exclusions>
        <exclusion>
            <artifactId>elasticsearch</artifactId>
            <groupId>org.elasticsearch</groupId>
        </exclusion>
    </exclusions>
</dependency>
```

一个简单的配置类 RestClient Config 实现代码如下：

```java
package com.lietu.demo.config;
import org.elasticsearch.client.RestHighLevelClient;
import org.springframework.context.annotation.Bean;
import org.springframework.context.annotation.Configuration;
import org.springframework.data.elasticsearch.client.ClientConfiguration;
import org.springframework.data.elasticsearch.client.RestClients;
import org.springframework.data.elasticsearch.config.AbstractElasticsearchConfiguration;

@Configuration
public class RestClientConfig extends AbstractElasticsearchConfiguration {
    @Override
    @Bean
    public RestHighLevelClient elasticsearchClient() {
        final ClientConfiguration clientConfiguration = ClientConfiguration.builder()
            .connectedTo("localhost:9200")
            .build();
        return RestClients.create(clientConfiguration).rest();
    }
}
```

下面为 RestHighLevelClient 创建一个具有 SSL 和基本身份验证配置的 Bean。

配置文件/src/main/resources/application.yml 的代码如下：

```yaml
elasticsearch:
  host: localhost
  port: 9200
  username: <username>
  password: <password>
```

Elastic Search Bean 配置，实现代码如下：

```java
@Configuration
```

```java
@EnableElasticsearchRepositories(basePackages = {"com.example.search"})
public class ESConfig extends AbstractElasticsearchConfiguration {
    @Value("${elasticsearch.host}")
    private String host;
    @Value("${elasticsearch.port:9200}")
    private int port;
    @Value("${elasticsearch.username}")
    private String username;
    @Value("${elasticsearch.password}")
    private String password;

    @Bean
    @Override
    public RestHighLevelClient elasticsearchClient() {
        ClientConfiguration.MaybeSecureClientConfigurationBuilder builder = ClientConfiguration.builder()
                .connectedTo(host+ ":" + port)
                .usingSsl()
                .withBasicAuth(username, password);
        final ClientConfiguration clientConfiguration = builder.build();
        return RestClients.create(clientConfiguration).rest();
    }
}
```

创建此配置后，Spring 将自动使用此 RestHighLevelClient 创建 Bean ElasticsearchOperations 的实例。

为自定义方法扩展 ElasticsearchRepository，代码如下：

```java
public interface BookRepository extends Repository<Book, String> {
    //通过名称和价格查找图书
    List<Book> findByNameAndPrice(String name, Integer price);
    //通过名称或者价格查找图书
    List<Book> findByNameOrPrice(String name, Integer price);
    //通过名称查找图书
    Page<Book> findByName(String name,Pageable page);
    //通过价格区间查找图书
    Page<Book> findByPriceBetween(int price,Pageable page);
    //通过名称模糊查找图书
    Page<Book> findByNameLike(String name,Pageable page);
    //通过内容查找图书
    @Query("{\"bool\" : {\"must\" : {\"term\" : {\"message\" : \"?0\"}}}}")
    Page<Book> findByMessage(String message, Pageable pageable);
}
```

使用 Repository 索引单个文档，代码如下：

```java
@Autowired
private SampleElasticsearchRepository repository;     //Elasticsearch 存储库
String documentId = "123456";                         //文档编号
SampleEntity sampleEntity = new SampleEntity();       //创建实体类
sampleEntity.setId(documentId);                       //设置文档编号
sampleEntity.setMessage("some message");              //设置消息
repository.save(sampleEntity);                        //保存实体类到存储库
```

使用 Repository 索引多个文档（批量索引），代码如下：

```
@Autowired
private SampleElasticsearchRepository repository;
//创建两个文档
String documentId = "123456";
SampleEntity sampleEntity1 = new SampleEntity();
sampleEntity1.setId(documentId);
sampleEntity1.setMessage("some message");
String documentId2 = "123457"
SampleEntity sampleEntity2 = new SampleEntity();
sampleEntity2.setId(documentId2);
sampleEntity2.setMessage("test message");
//实体类数组
List<SampleEntity> sampleEntities = Arrays.asList(sampleEntity1, sampleEntity2);
//批量索引
repository.save(sampleEntities);
```

ElasticsearchRestTemplate 是 Elasticsearch 操作的核心支持类。使用 ElasticsearchRestTemplate 索引单个文档，代码如下：

```
//创建单个文档
String documentId = "123456";
SampleEntity sampleEntity = new SampleEntity();
sampleEntity.setId(documentId);
sampleEntity.setMessage("some message");
IndexQuery indexQuery =new IndexQueryBuilder().withId(sampleEntity.getId()).withObject(sampleEntity).build();
elasticsearchTemplate.index(indexQuery);
```

使用 ElasticsearchTemplate 索引多个文档（批量索引），代码如下：

```
@Autowired
private ElasticsearchTemplate elasticsearchTemplate;
List<IndexQuery> indexQueries = new ArrayList<IndexQuery>();
//第一个文档
String documentId = "123456";
SampleEntity sampleEntity1 = new SampleEntity();
sampleEntity1.setId(documentId);
sampleEntity1.setMessage("some message");
IndexQuery indexQuery1 = new IndexQueryBuilder().withId(sampleEntity1.getId())
        .withObject(sampleEntity1).build();
indexQueries.add(indexQuery1);
//第二个文档
String documentId2 = "123457";
SampleEntity sampleEntity2 = new SampleEntity();
sampleEntity2.setId(documentId2);
sampleEntity2.setMessage("some message");
//构建索引查询
IndexQuery indexQuery2 = new IndexQueryBuilder().withId(sampleEntity2.getId())
        .withObject(sampleEntity2).build()
indexQueries.add(indexQuery2);
//批量索引
elasticsearchTemplate.bulkIndex(indexQueries);
```

使用 ElasticsearchTemplate 搜索实体，代码如下：

```
@Autowired
private ElasticsearchTemplate elasticsearchTemplate;
//构建查询类
SearchQuery searchQuery = new NativeSearchQueryBuilder()
    .withQuery(queryString(documentId).field("id"))
    .build();
//返回一页查询结果
Page<SampleEntity> sampleEntities =
    elasticsearchTemplate.queryForPage(searchQuery,SampleEntity.class);
```

使用实体类定义索引结构，代码如下：

```
import org.springframework.data.annotation.Id;
import org.springframework.data.annotation.Version;
import org.springframework.data.elasticsearch.annotations.Document;
import org.springframework.data.elasticsearch.annotations.Field;
import org.springframework.data.elasticsearch.annotations.FieldType;

//通过注解映射实体类到索引库
@Document(indexName = "book",type = "book" , shards = 1, replicas = 0, indexStoreType = "memory", refreshInterval = "-1")
public class Book {
    @Id
    private String id;            //图书编号
    private String name;
    private Long price;
    @Version
    private Long version;         //版本号
    public Map<Integer, Collection<String>> getBuckets() {
        return buckets;
    }
    public void setBuckets(Map<Integer, Collection<String>> buckets) {
        this.buckets = buckets;
    }
    //书对应的类别
    @Field(type = FieldType.Nested)
    private Map<Integer, Collection<String>> buckets = new HashMap();

    public Book(){…}

    public Book(String id, String name,Long version) {
        this.id = id;
        this.name = name;
        this.version = version;
    }
    //实体 Bean 需要的一些方法
    public String getId() {
        return id;
    }
    public void setId(String id) {
        this.id = id;
```

```
    }
    public String getName() {
        return name;
    }
    public void setName(String name) {
        this.name = name;
    }
    public Long getPrice() {
        return price;
    }
    public void setPrice(Long price) {
        this.price = price;
    }
    public long getVersion() {
        return version;
    }
    public void setVersion(long version) {
        this.version = version;
    }
}
```

实际执行索引结构定义的代码如下:

```
elasticsearchTemplate.putMapping(Book.class);
```

另一个使用注解的例子,代码如下:

```
@Entity(name="content")
@Document(indexName = "lietuim", type = "content")    //索引类型
public class ContentEntity {
    private int id;
    @Field(type = FieldType.text,searchAnalyzer = "standard", analyzer = "standard",store = true)
                                                //通过注解指定内容所用的分析器
    private String content1;
    @Field(type = FieldType.Double,store = true)    //指定要存储的浮点数
    private Double num1;
    @Field(type = FieldType.Date,store = true)    //日期类型的发布时间
    private Timestamp nowtime;
    @Field(type = FieldType.text,store = false)    //唯一编号
    private String uuid;
    @Field(type = FieldType.Integer,store = false)    //受欢迎程度
    private Integer likenum;
    private int cid;

    public ContentEntity(int id, String content1, Double num1,
        Timestamp nowtime, String uuid, Integer likenum, int cid) {    //构造方法
        this.id = id;
        this.content1 = content1;
        this.num1 = num1;
        this.nowtime = nowtime;
        this.uuid = uuid;
        this.likenum = likenum;
        this.cid = cid;
    }
```

```java
public ContentEntity() {
}

@Id
@GeneratedValue(strategy = IDENTITY)
@Column(name = "id", unique = true, nullable = false)        //生成的唯一编号
public int getId() {
    return id;
}
public void setId(int id) {
    this.id = id;
}

@Basic
@Column(name = "content1")
public String getContent1() {
    return content1;
}
public void setContent1(String content1) {
    this.content1 = content1;
}

@Basic
@Column(name = "num1")
public Double getNum1() {
    return num1;
}
public void setNum1(Double num1) {
    this.num1 = num1;
}

@Basic
@Column(name = "nowtime")
public Timestamp getNowtime() {
    return nowtime;
}
public void setNowtime(Timestamp nowtime) {
    this.nowtime = nowtime;
}

@Basic
@Column(name = "uuid")
public String getUuid() {
    return uuid;
}
public void setUuid(String uuid) {
    this.uuid = uuid;
}

@Basic
@Column(name = "likenum")                                    //评分
public Integer getLikenum() {
```

```
        return likenum;
    }
    public void setLikenum(Integer likenum) {
        this.likenum = likenum;
    }

    @Basic
    @Column(name = "cid")        //类别
    public int getCid() {
        return cid;
    }
    public void setCid(int cid) {
        this.cid = cid;
    }
}
```

5.3 实现自动完成

搜索输入框中的下拉提示，给用户一个有参考意义的搜索词表。本节使用 Spring Boot 实现自动完成的服务器端和客户端。

5.3.1 自动完成服务器端

当用户输入一个搜索字的同时，由 Rest 控制器从后台获取数据。我们先设计词典格式。它是由两列组成：第一列是词；第二列是搜索返回的结果数量。中间用冒号隔开，例如：

```
ball : 15
```

搜索词为 ball，搜索返回结果数量为 15。构造一个快速查找树 Ternary Search Trie 来实现这个词典，代码如下：

```
public class SuggestWordByTrie {
    private static Trie<SuggestItem> rootNode;
    private static SuggestWordByTrie swt =new SuggestWordByTrie();
    public static SuggestWordByTrie getInstantiation(){
        return swt;
    }

    public synchronized Trie<SuggestItem> getTrie() throws IOException{
        if(rootNode==null){
            rootNode =new Trie<SuggestItem>();
            initReader();
        }
        return rootNode;
    }

    public List<String> getTopWords(String prefix, int top) throws IOException{
        List<SuggestItem> list = getTrie().search(prefix);
        SuggestItem [] items =new SuggestItem[list.size()];
```

```
        list.toArray(items);
        QuickSelect.quickSelect(items,top);
        if(items.length<=top){
            top=items.length;
        }
        ArrayList<String> ret = new ArrayList<String>(top);
        for(int i=0;i<top;i++){
            ret.add(items[i].w);
        }

        return ret;
    }

    private void initReader() throws IOException {
        File f = new File(Thread.currentThread().getContextClassLoader().getResource(
                "static/dic/suggestword.txt").getFile());
        if(!f.exists()){
            return;
        }
        BufferedReader br =new BufferedReader(new FileReader(f));
        String line;
        while((line=br.readLine())!=null){
            String [] s = line.split(":");
            String word =s[0].trim();
            int count = Integer.parseInt(s[1].trim());
            rootNode.add(s[0].trim(), new SuggestItem(word,count));
        }
    }
}
```

为了在单独的控制类中实现自动完成功能,可以创建一个应用程序类,代码如下:

```
@SpringBootApplication
public class Application {
    public static void main(String[] args) {
        SpringApplication.run(Application.class, args);
    }
}
```

自动完成的 RestController 类可以改写成下面这样:

```
@RestController
public class AutoCompleteController {
    private static int top = 8;
    private static SuggestWordByTrie swt =SuggestWordByTrie.getInstantiation();

    @RequestMapping("/autocomplete")
    public List<SuggestItem> getAutoComplete(@RequestParam(name="q") String term)
throws IOException {
        return swt.getTop(term, top);
    }
}
```

执行以下命令测试 REST 端点 autocomplete。

```
>curl http://localhost/autocomplete?q=h
```

返回结果如下：

```
[{"c":6,"w":"hot"},{"c":5,"w":"horizontal"},{"c":5,"w":"headed"},{"c":6,"w":"hand
le"},{"c":6,"w":"hss"},{"c":6,"w":"hewlett"},{"c":8,"w":"headphone"},{"c":9,"w":"hand
held"}]
```

5.3.2 自动完成客户端

搜索首页 static/index.html 中的相关代码如下：

```
<form name="rootAction_index_Form" id="searchForm" method="get" action="./search/">
    <input type="text" name="query" maxlength="1000" size="50" value="" id="contentQuery" autocomplete="off">
    <button type="submit" name="search" id="searchButton">
        Search
    </button>
</form>
```

static/js/autocomplete.js 文件中的 JavaScript 代码如下：

```
/*
    function: addEvent

    @param: obj (Object)(Required)              //想要附加事件的对象

    @param: type (String)(Required)             //希望建立的事件类型

    @param: callback (Function)(Required)       //希望事件监听器调用的方法

    @param: eventReturn (Boolean)(Optional)     //是否要将事件对象返回给回调方法
*/
var addEvent = function(obj, type, callback, eventReturn) {
    if (obj == null || typeof obj === 'undefined')
        return;

    if (obj.addEventListener)
        obj.addEventListener(type, callback, eventReturn ? true : false);
    else
        obj["on" + type] = callback;
};

const keys = {
    ENTER: 13,
    ARROW_UP: 38,
    ARROW_DOWN: 40,
};

var suggestor = new function() {
    var listSelNum = 0;
    var listNum = 0;
    var inputText = "";

    //DOM 树增加 suggestorBox
```

```javascript
var boxElement = document.createElement('div');
boxElement.setAttribute("id", "suggestorBox");
boxElement.className = 'suggestorBox';
boxElement.style.display = "none";                    //默认不显示此元素
boxElement.style.position = "fixed";
boxElement.style.zIndex = "2";                        //设置堆叠顺序
boxElement.style.setProperty('background-color', 'rgb(255, 255, 255)');
boxElement.style.width = '278px';

document.body.appendChild(boxElement);

//增加 suggestorBox 的显示风格
var css = '#suggestorBox li:hover { background-color:rgba(82, 168, 236, 0.1); }';
var style = document.createElement('style');

if (style.styleSheet) {
    style.styleSheet.cssText = css;
} else {
    style.appendChild(document.createTextNode(css));
}

document.getElementsByTagName('head')[0].appendChild(style);

var olEle = document.createElement('ol');
olEle.style.cssText = 'list-style: none; padding: 0px; margin: 2px;'
boxElement.appendChild(olEle);

this.selectListUp = function selectListUp() {
    listSelNum--;
    if (listSelNum < 0) {
        listSelNum = listNum;
    }
}

this.selectListDown = function selectListDown() {
    listSelNum++;
    if (listSelNum > listNum) {
        listSelNum = 0;
    }
}

var textArea = document.getElementById('contentQuery');

var watchInput = function(evt) {
    suggestor.inputKeyDown(evt);
};

addEvent(textArea, "keydown", watchInput, true);

var searchForm = document.searchForm;

this.fixList = function() {
```

```javascript
        if (listSelNum > 0) {
            textArea.value =
                    boxElement.children[0].children[listSelNum - 1].innerHTML;
        }
        else {
            textArea.value = inputText;
        }
    }

    this.render = function() {
        for (k = 1; k <= listNum; k++) {
            if (k === listSelNum) {
                boxElement.children[0].children[k - 1].style
                        .setProperty('background-color', 'rgba(82, 168, 236, 0.1)');
            } else {
                boxElement.children[0].children[k - 1].style
                        .setProperty('background-color', 'rgb(255, 255, 255)');
            }
        }
    }

    var url = '/autoComplete?q=';

    this.wordsList = function() {
        boxElement.style.display = "none";
        //请求后端提供的自动完成接口
        fetch(url + textArea.value)
            .then(res => res.json())
            .then(res => {
                suggestor.createAutoCompleteList(res);
                suggestor.listNum = res.length;
            });
        boxElement.style.display = "block";
        listSelNum = 0;
    }

    this.inputKeyDown = function inputKeyDown(e) {
        if (e.keyCode === keys.ENTER) {
            document.getElementById('searchButton').click();
        }
        else if (e.keyCode === keys.ARROW_UP) {
            if (boxElement.style.display === "none") {
                return;
            }
            suggestor.selectListUp();
            suggestor.render();
            suggestor.fixList();
        } else if (e.keyCode === keys.ARROW_DOWN) {
            if (boxElement.style.display === "none") {
                return;
            }
            suggestor.selectListDown();
```

```
            suggestor.render();
            suggestor.fixList();
        }
        else if (
            (e.keyCode >= 48 && e.keyCode <= 90) ||
            (e.keyCode >= 96 && e.keyCode <= 105) ||
            (e.keyCode >= 186) ||
            e.keyCode === 8 ||
            e.keyCode === 32 ||
            e.keyCode === 46) {
            //使用在 setTimeout 函数内注册的函数来得到输入框中实际的值
            setTimeout(suggestor.wordsList, 1);
        }
    }

    var watchInputPress = function(evt) {
        suggestor.inputKeyPress(evt);
    };

    addEvent(textArea, "keyup", watchInputPress, true);

    this.inputKeyPress = function inputKeyPress(e) {
        if (
            (e.keyCode >= 48 && e.keyCode <= 90) ||
            (e.keyCode >= 96 && e.keyCode <= 105) ||
            (e.keyCode >= 186) ||
            e.keyCode === 8 ||
            e.keyCode === 32 ||
            e.keyCode === 46) {
            inputText = textArea.value;
        }
    }

    this.createAutoCompleteList = function(reslist) {
        olEle.innerHTML = '';

        for (k = 1; k <= reslist.length; k++) {
            var liEle = document.createElement('li');
            liEle.style.cssText = 'padding: 2px;';
            liEle.setAttribute("id", k);
            liEle.tabIndex = 0;
            liEle.innerHTML = reslist[k - 1];
            liEle.addEventListener("click", function(e) {
                textArea.value = this.innerHTML;
                searchForm.submit();
            });

            liEle.addEventListener("mouseout", function() {
                this.style.setProperty('background-color', 'rgb(255, 255, 255)');
            });

            liEle.addEventListener('mouseenter', function() {
```

```javascript
                    suggestor.listSelNum = this.id;
                    this.style.setProperty('background-color', 'rgba(82, 168, 236, 0.1)');
                }, true);

                liEle.addEventListener('keydown', function(e, me) {
                    if (e.keyCode === keys.ARROW_UP) {
                        suggestor.listSelNum = this.id;
                        suggestor.selectListUp();
                        this.style.setProperty('background-color', 'rgb(255, 255, 255)');
                        boxElement.children[0].children[suggestor.listSelNum - 1].style
                            .setProperty('background-color', 'rgba(82, 168, 236, 0.1)');
                        suggestor.fixList();
                        textArea.focus();
                    }
                    else if (e.keyCode === keys.ARROW_DOWN) {
                        suggestor.listSelNum = this.id;
                        suggestor.selectListDown();
                        this.style.setProperty('background-color', 'rgb(255, 255, 255)');
                        boxElement.children[0].children[suggestor.listSelNum - 1].style
                            .setProperty('background-color', 'rgba(82, 168, 236, 0.1)');
                        suggestor.fixList();
                        textArea.focus();
                    }
                }, true);
                olEle.appendChild(liEle);
            }
            listNum = reslist.length;
        }

        //匿名函数
        this.resize = function() {
            var pos = getPosition(textArea);
            boxElement.style.top = (pos.y + textArea.offsetHeight + 6) + 'px';
            boxElement.style.left = pos.x + 'px';
            boxElement.style.height = 'auto';
        }
    }

    suggestor.resize();

    var watchResize = function(evt) {
        suggestor.resize();
    };

    addEvent(window, 'resize', watchResize, true);

    //获取元素确切位置的辅助函数
    function getPosition(el) {
        var xPos = 0;
        var yPos = 0;
        while (el) {
            if (el.tagName == "BODY") {
```

```
            //获取BODY元素的确切位置
            var xScroll = el.scrollLeft || document.documentElement.scrollLeft;
            var yScroll = el.scrollTop || document.documentElement.scrollTop;

            xPos += (el.offsetLeft - xScroll + el.clientLeft);
            yPos += (el.offsetTop - yScroll + el.clientTop);
        } else {
            //对于所有非BODY元素获取确切位置
            xPos += (el.offsetLeft - el.scrollLeft + el.clientLeft);
            yPos += (el.offsetTop - el.scrollTop + el.clientTop);
        }

        el = el.offsetParent;
    }
    return {
        x: xPos,
        y: yPos
    };
}
```

在网页中引用 autocomplete.js，代码如下：

```
<script charset="utf-8" type="text/javascript" src="./js/autocomplete.js"></script>
```

suggestword.txt 可以放在 src\main\resources\static\dic\路径下。

5.4 界面国际化

为了将网页以不同语言呈现，可以使用 Spring Boot i18n 功能。这里的"i18n"是 internationalization（国际化）的简称，因为 internationalization 这个单词从字母 i 到字母 n 之间有 18 个英文字母而得名。

用一个消息创建两个属性文件：messages.properties 和 messages_zh.properties。messages.properties 文件的代码如下：

```
websiteTitle=lietuSearch
search="Search"
```

messages_zh.properties 文件的代码如下：

```
websiteTitle=猎兔搜索中文
search="搜索"
```

可以称这一组属性文件为一个资源包。Spring Boot i18n 自动配置扫描资源包并配置 org.springframework.context.MessageSource 实例。负责向 Freemarker 提供消息的 MessageResolver Method 类，其实现代码如下：

```
public class MessageResolverMethod implements TemplateMethodModelEx {
    private MessageSource messageSource;
    private Locale locale;
    public MessageResolverMethod(MessageSource messageSource, Locale locale) {
        this.messageSource = messageSource;
        this.locale = locale;
```

```
    }
    @Override
    public Object exec(List arguments) throws TemplateModelException {
        if (arguments.size() != 1) {
            throw new TemplateModelException("Wrong number of arguments");
        }
        String code = arguments.get(0).toString();
        if (code == null || code.isEmpty()) {
            throw new TemplateModelException("Invalid code value '" + code + "'");
        }
        return messageSource.getMessage(code, null, locale);
    }
}
```

创建一个简单的@RestController，它返回特定于语言环境的消息。Spring Boot 会自动配置 MessageSource Bean，代码如下：

```
@RestController
public class HelloController {
    private final Logger logger = LogManager.getLogger(HelloController.class);
    @Autowired
    MessageSource messageSource;
    @GetMapping("/")
    public String index(Locale locale) throws IOException, TemplateException {
        //传入的 Locale 对象代表了一个特定的地理、政治和文化地区
        logger.info("locale|" + locale.toString()); //输出 Locale 对象
        Configuration config = new Configuration(Configuration.VERSION_2_3_28);
        //配置设置
        TemplateLoader ctl = new ClassTemplateLoader(HelloController.class,
            "/templates/");
        TemplateLoader ftl1 = new FileTemplateLoader(new File(
            System.getProperty("user.dir")));
        MultiTemplateLoader mtl = new MultiTemplateLoader(
            new TemplateLoader[]{ftl1, ctl});
        config.setTemplateLoader(mtl);
        config.setDefaultEncoding("UTF-8");
        config.setTemplateExceptionHandler(
            TemplateExceptionHandler.RETHROW_HANDLER);
        //获取模板（注意,此处不涉及语言环境）
        String templateName = "index.ftl";
        Template template = config.getTemplate(templateName);

        Map<String, Object> modelMap = new HashMap<String, Object>();
        // Freemarker 模板中将提供名为"msg"的方法
        //语言环境在 msg()方法中发挥作用
        modelMap.put("msg", new MessageResolverMethod(messageSource, locale));

        //将 Freemarker 输出写入 StringWriter
        StringWriter stringWriter = new StringWriter();
        template.process(modelMap, stringWriter);
        //从 StringWriter 获取字符串
```

```
            String content = stringWriter.toString();
            return content;
    }
}
```

模板文件 index.ftl 的代码如下：

```
<html>
<head>
    <title>${msg("websiteTitle")}</title>
    <META http-equiv=Content-Type content="text/html; charset=utf-8">
    <link href="/CSS/default.css" rel="stylesheet" type="text/css" />
</head>
<body>
    <div id="sHeader">
        <div id="searcher">
            <form method="get" action="search">
                <input ID="tbKeyword" type="text" name="query" class="input_text" value="" />
                <input ID="btSearch" type=submit value=${msg("search")}/>
            </form>
        </div>
    </div>
</body>
</html>
```

5.5 本章小结

本章介绍了 Spring Boot 通过 Spring Data Solr 和 Spring Data Elasticsearch 整合搜索引擎，以及搜索界面的自动完成功能，然后介绍了 Spring Boot 对界面国际化的支持。

第 6 章

Web 应用程序开发

本章首先介绍前后端分离的 Web 应用程序开发，然后介绍如何使用 WebSocket 构建交互式 Web 应用程序。

6.1 使用 Bootstrap 实现搜索结果页面

使用前端框架 Bootstrap 实现搜索结果页面。首先下载 Bootstrap 分发包，执行命令如下：

```
#wget https://github.com/twbs/bootstrap/releases/download/v4.4.1/bootstrap-4.4.1-dist.zip
```

然后在网页的`<head></head>`标签中引入 Bootstrap 的样式库，代码如下：

```html
<head>
    <meta charset="UTF-8">
    <title>Bootstrap 模板</title>
    <meta name="viewport" content="width=device-width, initial-scale=1.0">
    <!-- 引入 Bootstrap -->
    <link href="https://maxcdn.bootstrapcdn.com/bootstrap/4.3.1/css/bootstrap.min.css" rel="stylesheet">
</head>
```

理论上来说，`<script></script>`标签放在 HTML 文档的任何位置都可以。为了快速加载网页，推荐将该标签放到`<body>`标签的末尾，包含到`<body></body>`标签内。

```html
<body>
    <h1>Hello, world!</h1>
    <!-- 包括所有已编译的插件 -->
    <script src="js/bootstrap.min.js"></script>
</body>
```

可以使用 CSS 预处理器 Sass（Syntactically Awesome Stylesheets）辅助 CSS 的开发。

安装预处理器 Sass，执行命令如下：

```
>choco install sass
```

使用 Sass，执行命令如下：

```
>sass styles.scss styles.css
```

6.2 重试

Spring Retry 是一款支持失败后重试操作的框架。为了使用 Spring Retry，需在 build.gradle 文件中增加如下依赖项。

```
ext {
    springVersion= "5.1.5.RELEASE"
}

dependencies {
    compile group: 'org.springframework', name: 'spring-aspects', version: springVersion
    compile group: 'org.springframework', name: 'spring-aop', version: springVersion
    compile group: 'org.springframework.retry', name: 'spring-retry', version: '1.2.4.RELEASE'
    compile group: 'org.springframework', name: 'spring-core', version: springVersion
    compile group: 'org.springframework', name: 'spring-context', version: springVersion

    compile group: 'com.squareup.okhttp3', name: 'okhttp', version: '3.13.1'
}
```

首先定义一个需要重试的服务类，然后在配置类中提供得到这个服务类的实例方法。下载服务类 DownService 的实现代码如下：

```
@Component
public class DownService {
    public static int i = 0;   //下载次数
    @Retryable( value = {Exception.class},maxAttempts = 5000)
    public String getContent(String url, OkHttpClient client) throws Exception {
        System.out.println("Down Attempt:" + i++);
        okhttp3.Request.Builder rb = new Request.Builder().url(url);
        Request request = rb.build();
        Response response = client.newCall(request).execute();

        int responseCode = response.code();
        String contentType = response.header("Content-Type");
        if (responseCode != 200 || contentType == null) {
            throw new IOException("responseCode " + responseCode);
        }
        return response.body().string();
    }
}
```

CrawlerApplication 应用程序类调用服务类实现重试下载，代码如下：

```
@Configuration
@EnableRetry
public class CrawlerApplication {
    public static void main(String[] args) throws Exception {
        ConfigurableApplicationContext context = new AnnotationConfigApplicationContext(
            CrawlerApplication.class);
        String url = "http://www.lietu.com";
        OkHttpClient client = new OkHttpClient.Builder().connectTimeout(1000, TimeUnit.SECONDS)
```

```
                .readTimeout(200, TimeUnit.SECONDS)
                .retryOnConnectionFailure(true).build();
        String html = context.getBean(DownService.class).getDoc(url, client);
        System.out.println(html);
    }

    @Bean
    public DownService service() {
        return new DownService();
    }
}
```

为了改进 Spring Retry，可以下载 Spring Retry 源代码，然后在本地修改并编译源代码。可以使用以下 git 命令下载源代码：

```
>git clone https://github.com/spring-projects/spring-retry.git
```

也可以使用以下 svn 命令下载源代码：

```
>svn export https://github.com/spring-projects/spring-retry/trunk/
```

使用 Maven 编译源代码，执行命令如下：

```
>mvn clean install
```

为了忽略编译过程中的错误 MavenReportException: Error while generating Javadoc，可以修改 pom.xml 文件，为 Javadoc 插件增加配置项<failOnError>false</failOnError>，代码如下：

```xml
<plugin>
    <artifactId>maven-javadoc-plugin</artifactId>
    <executions>
        <execution>
            <id>javadoc</id>
            <goals>
                <goal>jar</goal>
            </goals>
            <phase>package</phase>
            <configuration>
                <failOnError>false</failOnError>
            </configuration>
        </execution>
    </executions>
</plugin>
```

创建一个项目测试打包出来的 spring-retry。这个项目的 build.gradle 文件设置成从本地库加载依赖项，代码如下：

```
repositories {
    mavenLocal()
    jcenter()
}

dependencies {
    compile 'org.springframework:spring-context:4.3.22.RELEASE'
    compile 'org.springframework:spring-core:4.3.22.RELEASE'
    compile 'javax.annotation:javax.annotation-api:1.3.2'
    compile 'org.aspectj:aspectjweaver:1.8.9'
```

```
        compile 'log4j:log4j:1.2.17'
        compile group: 'org.springframework.retry', name: 'spring-retry', version: '1.3.0.
BUILD-SNAPSHOT'
        compile group: 'javax.annotation', name: 'javax.annotation-api', version: '1.3.2'
        testCompile 'junit:junit:4.12'
        testCompile 'org.easymock:easymock:2.3'
        testCompile 'org.mockito:mockito-core:1.10.19'
    }
    sourceCompatibility = '1.8'
```

测试支持重试的服务，代码如下：

```
public class App {
    public static void main(String[] args) {
        AnnotationConfigApplicationContext context =
            new AnnotationConfigApplicationContext(TestConfiguration.class);
        RetryableService service = context.getBean(RetryableService.class);
        service.service();
        System.out.println("Count "+service.getCount());
        context.close();
    }

    @Configuration
    @EnableRetry
    protected static class TestConfiguration {
        @Bean
        public RetryableService retryable() {        //配置一个支持重试的服务
            return new RetryableService();
        }
    }

    @Retryable(maxAttempts = 5000)
    protected static class RetryableService {
        private int count = 0;
        public void service() {
            System.out.println(count);
            if (count++ < 2) {
                throw new RuntimeException("Planned");
            }
        }

        public int getCount() {
            return count;
        }
    }
}
```

6.3 整合 Kafka

当耗时操作影响了性能时，如果允许异步处理，可以选择消息队列进行异步处理。Kafka

是一个开源的分布式消息队列系统。本节首先介绍安装 Kafka，然后介绍创建一个能够连接给定 Apache Kafka 代理实例的 Spring Boot 应用程序。另外，介绍从 Kafka 主题中产生和消费消息。

要安装 Kafka，必须先在系统上安装 Java。然后为 Kafka 设置 ZooKeeper，ZooKeeper 负责管理 Kafka 集群状态。Apache ZooKeeper 是 Apache 软件基金会的软件项目，它本质上是为分布式系统提供分层键值存储的服务，用于为大型分布式系统提供分布式配置服务、同步服务和命名注册表。

ZooKeeper 的设置流程如下。

（1）从网址 https://www.apache.org/dyn/closer.cgi/zookeeper/ 下载 ZooKeeper。

（2）解压缩文件。在 conf 目录中，重命名文件 zoo_sample.cfg 为 zoo.cfg。

（3）zoo.cfg 文件保留 ZooKeeper 的配置，即 ZooKeeper 实例将在哪个端口监听，以及数据目录等。

（4）默认的监听端口为 2181。可以通过更改 zoo.cfg 中的 clientPort 选项来更改此端口。

（5）默认数据目录为 /tmp/data。更改此设置，因为不希望在某个随机时间段后删除 ZooKeeper 的数据。在 ZooKeeper 目录中创建一个名为 data 的文件夹，并在 zoo.cfg 中更改 dataDir 选项。

（6）进入 bin 目录。

（7）通过执行命令 ./zkServer.sh start 来启动 ZooKeeper。而通过执行命令 ./zkServer.sh stop 则可以停止 ZooKeeper。

Kafka 的设置流程如下。

（1）从 https://kafka.apache.org/downloads 下载 Kafka 的稳定版本。

（2）解压缩文件 kafka_2.12-2.5.0.tgz。Kafka 实例（代理）配置保留在 config 目录中。

（3）进入 config 目录，打开文件 server.properties。

（4）从 listeners 属性中删除注释，即 listeners = PLAINTEXT://:9092，则 Kafka 代理将监听端口 9092。

（5）将 log.dirs 更改为 /kafka_home_directory/kafka-logs。

（6）检查 zookeeper.connect 属性并根据需要进行更改。Kafka 代理将连接到该 ZooKeeper 实例。

（7）转到 Kafka 主目录并执行命令 ./bin/kafka-server-start.sh config/server.properties。通过执行命令 ./bin/kafka-server-stop.sh 可以停止 Kafka 代理。

打开 Spring Initializr 并创建具有以下依赖项的 Spring Boot 应用程序。

- Spring for Apache Kafka。
- Spring Web。

生成的项目在 pom.xml 中具有以下依赖项。

```
<dependencies>
    <dependency>
        <groupId>org.springframework.boot</groupId>
        <artifactId>spring-boot-starter-web</artifactId>
    </dependency>
```

```xml
        <dependency>
            <groupId>org.springframework.kafka</groupId>
            <artifactId>spring-kafka</artifactId>
        </dependency>
        <dependency>
            <groupId>org.springframework.boot</groupId>
            <artifactId>spring-boot-starter-test</artifactId>
            <scope>test</scope>
            <exclusions>
                <exclusion>
                    <groupId>org.junit.vintage</groupId>
                    <artifactId>junit-vintage-engine</artifactId>
                </exclusion>
            </exclusions>
        </dependency>
        <dependency>
            <groupId>org.springframework.kafka</groupId>
            <artifactId>spring-kafka-test</artifactId>
            <scope>test</scope>
        </dependency>
</dependencies>
```

在首选的 IDE 中导入项目。在 application.yaml 文件中，添加 Kafka 代理地址及与消费者和生产者相关的配置，代码如下：

```yaml
server:
  port: 9000
spring:
  kafka:
    consumer:
      bootstrap-servers: localhost:9092
      group-id: group-id
      auto-offset-reset: earliest
      key-deserializer: org.apache.kafka.common.serialization.StringDeserializer
      value-deserializer: org.apache.kafka.common.serialization.StringDeserializer
    producer:
      bootstrap-servers: localhost:9092
      key-deserializer: org.apache.kafka.common.serialization.StringDeserializer
      value-deserializer: org.apache.kafka.common.serialization.StringDeserializer
```

KafKaProducerService 类使用自动装配的 KafkaTemplate 将消息发送到已配置的主题名称。KafKaConsumerService 类使用@KafkaListener 来接收配置了主题名称的消息。

KafKaProducerService.java 实现代码如下：

```java
import org.slf4j.Logger;
import org.slf4j.LoggerFactory;
import org.springframework.beans.factory.annotation.Autowired;
import org.springframework.kafka.core.KafkaTemplate;
import org.springframework.stereotype.Service;
import com.howtodoinjava.kafka.demo.common.AppConstants;

@Service
public class KafKaProducerService
```

```java
{
    private static final Logger logger = LoggerFactory.getLogger(KafKaProducerService.class);
    @Autowired
    private KafkaTemplate<String, String> kafkaTemplate;
    public void sendMessage(String message)
    {
        logger.info(String.format("Message sent -> %s", message));
        this.kafkaTemplate.send(AppConstants.TOPIC_NAME, message);
    }
}
```

KafKaConsumerService.java 实现代码如下：

```java
import org.slf4j.Logger;
import org.slf4j.LoggerFactory;
import org.springframework.kafka.annotation.KafkaListener;
import org.springframework.stereotype.Service;
import com.howtodoinjava.kafka.demo.common.AppConstants;

@Service
public class KafKaConsumerService
{
    private final Logger logger = LoggerFactory.getLogger(KafKaConsumerService.class);

    @KafkaListener(topics = AppConstants.TOPIC_NAME,
            groupId = AppConstants.GROUP_ID)
    public void consume(String message)
    {
        logger.info(String.format("Message recieved -> %s", message));
    }
}
```

AppConstants.java 实现代码如下：

```java
public class AppConstants
{
    public static final String TOPIC_NAME = "test";
    public static final String GROUP_ID = "group_id";
}
```

控制器负责使用 RESTful API 从用户获取消息，并将消息移交给生产者服务以将其发布到 Kafka 主题。

KafkaProducerController.java 实现代码如下：

```java
import org.springframework.beans.factory.annotation.Autowired;
import org.springframework.web.bind.annotation.PostMapping;
import org.springframework.web.bind.annotation.RequestMapping;
import org.springframework.web.bind.annotation.RequestParam;
import org.springframework.web.bind.annotation.RestController;
import com.howtodoinjava.kafka.demo.service.KafKaProducerService;

@RestController
@RequestMapping(value = "/kafka")
public class KafkaProducerController
```

```
{
    private final KafKaProducerService producerService;

    @Autowired
    public KafkaProducerController(KafKaProducerService producerService)
    {
        this.producerService = producerService;
    }

    @PostMapping(value = "/publish")
    public void sendMessageToKafkaTopic(@RequestParam("message") String message)
    {
        this.producerService.sendMessage(message);
    }
}
```

使用 RESTful API 测试器，并在查询参数"message"中向 API http://localhost:9000/kafka/publish 发布一些消息。例如，发布 Post 消息：

```
http://localhost:9000/kafka/publish?message=Alphabet
```

观察控制台日志：

```
2020-05-24 23:36:47.132  INFO 2092 --- [nio-9000-exec-4]
c.h.k.demo.service.KafKaProducerService  : Message sent -> Alphabet

2020-05-24 23:36:47.138  INFO 2092 --- [ntainer#0-0-C-1]
c.h.k.demo.service.KafKaConsumerService  : Message recieved -> Alphabet
```

如果已经在命令提示符下打开了 Kafka 控制台消费者，也会看到该消息。

6.4 测试

使用 JUnit 来测试页面。当选用 Gradle 构建项目时，可以将支持测试的 build.gradle 文件修改成如下代码。

```
plugins {
    id 'org.springframework.boot' version '2.2.2.RELEASE'
    id 'io.spring.dependency-management' version '1.0.8.RELEASE'
    id 'java'
}
group = 'com.example'
version = '0.0.1-SNAPSHOT'
sourceCompatibility = '1.8'
repositories {
    maven {url 'http://maven.aliyun.com/nexus/content/groups/public/'}
    jcenter()
}
dependencies {
    implementation 'org.springframework.boot:spring-boot-starter-web'
    testImplementation('org.springframework.boot:spring-boot-starter-test') {
        exclude group: 'org.junit.vintage', module: 'junit-vintage-engine'
```

```
    }
}
test {
    useJUnitPlatform()
}
```

@SpringBootTest 注解负责告知 Spring Boot 寻找一个主要的配置类，并使用该类来启动 Spring Boot 应用程序上下文。使用@SpringBootTest 注解测试控制类，代码如下：

```
@RunWith(SpringRunner.class)
@SpringBootTest(classes = SearchSolr.class)        //引入 Spring Boot 应用程序上下文
public class ControllerTest {
    @Autowired
    private SearchSolr searchController;            //自动注入
    @Test
    public void testControllerExists() {
        Assert.assertNotNull(searchController);
    }
}
```

使用 MockMvc 实现不启动服务器测试 HTTP 请求，代码如下：

```
import org.junit.jupiter.api.Test;
import org.springframework.beans.factory.annotation.Autowired;
import org.springframework.boot.test.autoconfigure.web.servlet.AutoConfigureMockMvc;
import org.springframework.boot.test.context.SpringBootTest;
import org.springframework.test.web.servlet.MockMvc;
import static org.springframework.test.web.servlet.request.MockMvcRequestBuilders.get;
import static org.springframework.test.web.servlet.result.MockMvcResultHandlers.print;
import static org.springframework.test.web.servlet.result.MockMvcResultMatchers.status;

@SpringBootTest
@AutoConfigureMockMvc
public class TestingWebApplicationTest {
    @Autowired
    private MockMvc mockMvc;
    @Test
    public void shouldReturnDefaultMessage() throws Exception {
        this.mockMvc.perform(get("/search?query=test")).andDo(print()).andExpect(status().isOk());
    }
}
```

界面国际化相关代码参见 5.4 节，这里不再详述。

6.5 React 框架实现前后端分离的 Web 应用程序

本节介绍将 Create React App 与 Spring Boot 结合使用来创建前后端分离的搜索界面。

首先，使用 https://start.spring.io 创建一个 Spring Boot 项目。添加 Web 依赖项，将 groupId 设置为 com.lietu，将 artifactId 设置为 spring-and-react。

现在，应该有一个看起来像以下这样带有 POM 的项目。

```xml
<?xml version="1.0" encoding="UTF-8"?>
<project xmlns="http://maven.apache.org/POM/4.0.0" xmlns:xsi="http://www.w3.org/2001/XMLSchema-instance"
    xsi:schemaLocation="http://maven.apache.org/POM/4.0.0 http://maven.apache.org/xsd/maven-4.0.0.xsd">
    <modelVersion>4.0.0</modelVersion>

    <groupId>com.lietu</groupId>
    <artifactId>spring-and-react</artifactId>
    <version>0.0.1-SNAPSHOT</version>
    <packaging>jar</packaging>

    <name>spring-and-react</name>
    <description>Demo project for Spring Boot</description>

    <parent>
        <groupId>org.springframework.boot</groupId>
        <artifactId>spring-boot-starter-parent</artifactId>
        <version>2.0.1.RELEASE</version>
        <relativePath/> <!-- lookup parent from repository -->
    </parent>

    <properties>
        <project.build.sourceEncoding>UTF-8</project.build.sourceEncoding>
        <project.reporting.outputEncoding>UTF-8</project.reporting.outputEncoding>
        <java.version>1.8</java.version>
    </properties>

    <dependencies>
        <dependency>
            <groupId>org.springframework.boot</groupId>
            <artifactId>spring-boot-starter-web</artifactId>
        </dependency>
        <dependency>
            <groupId>org.springframework.boot</groupId>
            <artifactId>spring-boot-starter-test</artifactId>
            <scope>test</scope>
        </dependency>
    </dependencies>

    <build>
        <plugins>
            <plugin>
                <groupId>org.springframework.boot</groupId>
                <artifactId>spring-boot-maven-plugin</artifactId>
            </plugin>
        </plugins>
    </build>
</project>
```

在 pom.xml 文件中添加 SolrJ 和 Spring Data Solr 依赖项，代码如下：

```xml
<dependency>
    <groupId>org.apache.solr</groupId>
    <artifactId>solr-solrj</artifactId>
    <version>8.2.0</version>
</dependency>
<dependency>
    <groupId>org.springframework.data</groupId>
    <artifactId>spring-data-solr</artifactId>
    <version>4.0.5.RELEASE</version>
</dependency>
```

添加一个名为 com.lietu.springandreact.SearchController 的控制器类，代码如下：

```java
@RestController
public class SearchController {
    @Autowired
    private SolrConfig configuration;
    @RequestMapping("/api/search")
    public List<String> getAutoComplete(@RequestParam(name="q") String term) throws Exception {
        SolrClient client = configuration.solrClient();
        String queryStr = "name:" + term;
        SolrQuery query = new SolrQuery(queryStr);
        QueryResponse response = client.query(query);
        ArrayList<String> ret = new ArrayList<String>();
        for (SolrDocument d : response.getResults()) {
            String name = (String) d.getFieldValue("name");
            ret.add(name);
        }
        return ret;
    }
}
```

在一个命令行窗口中，使用以下命令启动应用程序。

```
>mvn spring-boot:run
```

上面的命令运行了一个 Spring Boot 模块。一个模块是可以与其他模块的代码库分开维护和构建的代码库。

在另一个窗口中，使用 curl 命令或 Web 浏览器获取 http://localhost:8080/api/search?q = test。

```
$ curl http://localhost:8080/api/search?q=test
```

返回结果如下：

```
["Test with some GB18030 encoded characters","Test with some UTF-8 encoded characters"]
```

安装 Node.js，然后使用工具 npx 创建前端项目，执行命令如下：

```
$ npx create-react-app frontend
```

启动这个项目，执行命令如下：

```
$ cd frontend
$ npm start
```

在 Spring Boot 中有一个运行在本地主机的 8080 端口（http://localhost:8080）的后端服务器，在 React 中有一个运行在本地主机的 3000 端口（http://localhost:3000）的前端服务器。现希望能够调用在后端的服务,并在前端显示结果。为了做到这一点,要求前端服务器代理从:3000 到:8080 的任何请求。

将代理条目添加到 frontend/package.json,这将确保位于:3000 的 Web 服务器代理到 http://localhost:3000/api/*的任何请求到 http://localhost:8080/api。同时,这将使我们能够调用后端而不会遇到任何跨域资源共享(CORS)问题。

frontend/package.json 文件代码如下:

```json
{
  "name": "frontend",
  "version": "0.1.0",
  "private": true,
  "dependencies": {
    "react": "^16.12.0",
    "react-dom": "^16.12.0",
    "react-scripts": "^3.3.0"
  },
  "scripts": {
    "start": "react-scripts start",
    "build": "react-scripts build",
    "test": "react-scripts test --env=jsdom",
    "eject": "react-scripts eject"
  },
  "proxy": "http://localhost:8080",
  "browserslist": {
    "production": [
      ">0.2%",
      "not dead",
      "not op_mini all"
    ],
    "development": [
      "last 1 chrome version",
      "last 1 firefox version",
      "last 1 safari version"
    ]
  }
}
```

接下来,在前端添加一个 rest 调用。frontend/src/App.js 文件代码如下:

```
import React, {Component, useState, useEffect} from 'react';
import logo from './logo.svg';
import './App.css';

function App() {
  const [keyword, setKeyword] = React.useState("");
  const [message, setMessage] = React.useState("");
  const handleSubmit = (e: React.FormEvent) => {
    e.preventDefault();
    var url = '/api/search?q=';
    //请求后端提供的自动完成接口
    fetch(url + keyword)
```

```
          .then(response => response.text())
          .then(message => {
            setMessage(message);
          });
  }
  return (
    <div className="App">
      <form onSubmit={handleSubmit}>
        <div>
          <label htmlFor="keyword">keyword</label>
          <input
            id="keyword"
            value={keyword}
            onChange={(e) => setKeyword(e.target.value)}
          />
        </div>
        <button>Search</button>
      </form>
      {message}
    </div>
  );
}
export default App;
```

确保后端正在运行，然后重新启动前端。现在，应该能够通过位于 http://localhost:3000/ 的前端服务器获取搜索服务。

6.6 使用 WebSocket 构建交互式 Web 应用程序

WebSocket 是从 HTML 5 开始提供的一种浏览器与服务器间进行全双工通信的网络技术。依靠 WebSocket 可以实现客户端和服务器端的长连接、双向实时通信。

WebSocket 的工作流程是：

（1）浏览器通过 JavaScript 向服务端发出建立 WebSocket 连接的请求；

（2）在 WebSocket 连接建立成功后，客户端和服务端就可以通过 TCP 连接传输数据。

在 Maven 中，需要 Spring Boot WebSocket 依赖项。pom.xml 文件代码如下：

```
<?xml version="1.0" encoding="UTF-8"?>
<project xmlns="http://maven.apache.org/POM/4.0.0" xmlns:xsi="http://www.w3.org/2001/XMLSchema-instance"
         xsi:schemaLocation="http://maven.apache.org/POM/4.0.0 http://maven.apache.org/xsd/maven-4.0.0.xsd">
    <modelVersion>4.0.0</modelVersion>
    <groupId>com.lietu</groupId>
    <artifactId>boot-websocket</artifactId>
    <version>1.0-SNAPSHOT</version>

    <parent>
        <groupId>org.springframework.boot</groupId>
```

```xml
        <artifactId>spring-boot-starter-parent</artifactId>
        <version>2.2.5.RELEASE</version>
    </parent>

    <dependencies>
        <dependency>
            <groupId>org.springframework.boot</groupId>
            <artifactId>spring-boot-starter-websocket</artifactId>
        </dependency>
        <dependency>
            <groupId>org.json</groupId>
            <artifactId>json</artifactId>
            <version>20171018</version>
        </dependency>
    </dependencies>

    <build>
        <plugins>
            <plugin>
                <groupId>org.springframework.boot</groupId>
                <artifactId>spring-boot-maven-plugin</artifactId>
            </plugin>
        </plugins>
    </build>

    <repositories>
      <repository>
        <id>aliyun</id>
        <url>http://maven.aliyun.com/nexus/content/groups/public/</url>
      </repository>
    </repositories>

</project>
```

创建如下的 Spring Boot 引导类：

```
package com.lietu.websocket.config;
import org.springframework.boot.SpringApplication;
import org.springframework.boot.autoconfigure.SpringBootApplication;

@SpringBootApplication
public class Application {
    public static void main(String[] args) {
        SpringApplication.run(Application.class, args);
    }
}
```

在服务器端，接收数据并回复给客户端。在 Spring Boot 中，可以使用 TextWebSocketHandler 或 BinaryWebSocketHandler 创建自定义处理程序。其中，BinaryWebSocketHandler 用于处理图像等类型更丰富的数据。在下面的例子中，由于只需要处理文本，因此选用 TextWebSocketHandler。

```
package com.lietu.websocket.config;
import java.io.IOException;
import org.json.JSONObject;
```

```java
import org.springframework.stereotype.Component;
import org.springframework.web.socket.TextMessage;
import org.springframework.web.socket.WebSocketSession;
import org.springframework.web.socket.handler.TextWebSocketHandler;
@Component
public class SocketTextHandler extends TextWebSocketHandler {
    @Override
    public void handleTextMessage(WebSocketSession session, TextMessage message)
            throws InterruptedException, IOException {
        String payload = message.getPayload();
        JSONObject jsonObject = new JSONObject(payload);
        session.sendMessage(new TextMessage("Hi " + jsonObject.get("user") + " how may we help you?"));
    }
}
```

为了告知 Spring Boot 将客户端请求转发到端点，需要注册处理程序，代码如下：

```java
package com.lietu.websocket.config;
import org.springframework.context.annotation.Configuration;
import org.springframework.web.socket.config.annotation.EnableWebSocket;
import org.springframework.web.socket.config.annotation.WebSocketConfigurer;
import org.springframework.web.socket.config.annotation.WebSocketHandlerRegistry;

@Configuration
@EnableWebSocket
public class WebSocketConfig implements WebSocketConfigurer {
    public void registerWebSocketHandlers(WebSocketHandlerRegistry registry) {
        registry.addHandler(new SocketTextHandler(), "/user");
    }
}
```

接下来，实现用于建立 WebSocket 和进行调用的 UI 部分，app.js 文件代码如下：

```javascript
var ws;
function setConnected(connected) {
    $("#connect").prop("disabled", connected);
    $("#disconnect").prop("disabled", !connected);
}
function connect() {
    ws = new WebSocket('ws://localhost:8080/user');
    ws.onmessage = function(data) {
        helloWorld(data.data);
    }
    setConnected(true);
}
function disconnect() {
    if (ws != null) {
        ws.close();
    }
    setConnected(false);
    console.log("Websocket is in disconnected state");
}
function sendData() {
```

```javascript
    var data = JSON.stringify({
        'user' : $("#user").val()
    })
    ws.send(data);
}
function helloWorld(message) {
    $("#helloworldmessage").append(" " + message + "");
}

$(function() {
    $("form").on('submit', function(e) {
        e.preventDefault();
    });
    $("#connect").click(function() {
        connect();
    });
    $("#disconnect").click(function() {
        disconnect();
    });
    $("#send").click(function() {
        sendData();
    });
});
```

index.html 文件代码如下：

```html
<!DOCTYPE html>
<html>
<head>
    <title>WebSocket Chat Application </title>
    <link href="/bootstrap.min.css" rel="stylesheet">
    <link href="/style.css" rel="stylesheet">
    <script src="/jquery-1.10.2.min.js"></script>
    <script src="/app.js"></script>
</head>
<body>
<div id="main-content" class="container">
    <div class="row">
        <div class="col-md-8">
            <form class="form-inline">
                <div class="form-group">
                    <label for="connect">Chat Application:</label>
                    <button id="connect" type="button">Start New Chat</button>
                    <button id="disconnect" type="button" disabled="disabled">End Chat
                    </button>
                </div>
            </form>
        </div>
    </div>
    <div class="row">
        <div class="col-md-12">
            <table id="chat">
                <thead>
```

```html
            <tr>
                <th>Welcome user. Please enter you name</th>
            </tr>
            </thead>
            <tbody id="helloworldmessage">
            </tbody>
          </table>
        </div>
        <div class="row">
          <div class="col-md-6">
            <form class="form-inline">
              <div class="form-group">
                <textarea id="user" placeholder="Write your message here…" required></textarea>
              </div>
              <button id="send" type="submit">Send</button>
            </form>
          </div>
        </div>
      </div>
    </div>
  </body>
</html>
```

启动应用程序，转到本地主机的 8080 端口（http://localhost:8080）单击开始新的会话，将会打开 WebSocket 连接。

6.7　本章小结

本章介绍了利用 Bootstrap 和 FreeMarker 模板引擎实现搜索界面，以及使用 Spring Boot 实现界面模板国际化（参见 5.4 节）的方法、搜索输入框自动完成功能的实现方法。此外，还介绍了前端框架 React 结合 Spring Boot 后端实现搜索界面，并举例介绍了使用 WebSocket 构建一个简单的对话应用程序。

第 7 章

监控 Spring Boot 应用程序

本章首先介绍使用 Spring Boot Actuator 监控 Spring Boot 应用程序，然后介绍使用 Elastic 栈处理应用程序日志。

7.1 Spring Boot Actuator

Spring Boot Actuator 是 Spring Boot Framework 的子项目，可以使用它来帮助我们监控和管理 Spring Boot 应用程序。Spring Boot Actuator 包含执行器端点（资源所在的位置），我们可以使用 HTTP 和 JMX 端点来管理和监控 Spring Boot 应用程序。如果要在应用程序中获得生产就绪的功能，则应使用 Spring Boot Actuator。

下面通过一个例子来了解执行器的概念。

打开 Spring Initializr 并创建一个 Maven 项目，添加以下依赖项：Spring Web、Spring Boot Starter Actuator 和 Spring Data Rest HAL Browser，提供工件 ID 为 spring-boot-actuator-example。

生成项目后，创建一个名为 DemoRestController 的控制器类，代码如下：

```
package com.javatpoint;
import org.springframework.web.bind.annotation.GetMapping;
import org.springframework.web.bind.annotation.RestController;

@RestController
public class DemoRestController {
    @GetMapping("/hello")
    public String hello() {
        return "Hello User!";
    }
}
```

打开 application.properties 文件并通过添加以下语句禁用执行器的安全性功能。

```
management.security.enabled=false
```

运行 SpringBootActuatorExampleApplication.java 文件。打开浏览器并调用网址 http://localhost:8080/actuator，返回信息如下：

```
{"_links":{"self":{"href":"http://localhost:8080/actuator","templated":false},
"health":{"href":"http://localhost:8080/actuator/health","templated":false},"health-
path":{"href":"http://localhost:8080/actuator/health/{*path}","templated":true},"info":
{"href":"http://localhost:8080/actuator/info","templated":false}}}
```

默认情况下，该应用程序在端口 8080 上运行。执行器启动后，我们可以看到通过 HTTP 公开的所有端点列表。可以通过调用网址 http://localhost:8080/actuator/health 来调用运行状况端点。该端点表示状态为 UP。这意味着该应用程序运行状况良好且正在运行，不会中断。同样，可以调用其他端点来帮助监视和管理 Spring Boot 应用程序。

7.2 Elastic 栈日志监控

微服务之间是相互隔离的，它们不共享公共数据库和日志文件。随着微服务数量的增加及使用自动连续集成工具启用云部署，当我们遇到任何问题时，非常有必要提供一些调试组件等的准备。

Elastic 栈用于实时搜索、分析和可视化日志数据。Elastic 栈包括以下几个组件。
- logstash：用于日志结构化、标签化，支持 DSL 方式将数据进行结构化。
- Elasticsearch：用于提供日志相关的索引，使得日志能够有效地检索。
- kibana：用于提供日志检索、特定 metric 展示的面板、方便使用的 UI。

首先创建一个 Spring Boot 项目，然后添加一个 RestController 类。该类将公开一些端点，例如/elk、/elkdemo、/exception。

```java
package com.example.howtodoinjava.elkexamplespringboot;
import java.io.PrintWriter;
import java.io.StringWriter;
import java.util.Date;
import org.apache.log4j.Level;
import org.apache.log4j.Logger;
import org.springframework.beans.factory.annotation.Autowired;
import org.springframework.boot.SpringApplication;
import org.springframework.boot.autoconfigure.SpringBootApplication;
import org.springframework.context.annotation.Bean;
import org.springframework.core.ParameterizedTypeReference;
import org.springframework.http.HttpMethod;
import org.springframework.web.bind.annotation.RequestMapping;
import org.springframework.web.bind.annotation.RestController;
import org.springframework.web.client.RestTemplate;

@SpringBootApplication
public class ElkExampleSpringBootApplication {
    public static void main(String[] args) {
        SpringApplication.run(ElkExampleSpringBootApplication.class, args);
    }
}

@RestController
class ELKController {
    private static final Logger LOG = Logger.getLogger(ELKController.class.getName());
    @Autowired
    RestTemplate restTemplete;
```

```
    @Bean
    RestTemplate restTemplate() {
        return new RestTemplate();
    }
    @RequestMapping(value = "/elkdemo")
    public String helloWorld() {
        String response = "Hello user ! " + new Date();
        LOG.log(Level.INFO, "/elkdemo - &gt; " + response);
        return response;
    }

    @RequestMapping(value = "/elk")
    public String helloWorld1() {
        String response = restTemplete.exchange("http://localhost:8080/elkdemo",
HttpMethod.GET, null, new ParameterizedTypeReference() {
        }).getBody();
        LOG.log(Level.INFO, "/elk - &gt; " + response);
        try {
            String exceptionrsp = restTemplete.exchange("http://localhost:8080/exception",
HttpMethod.GET, null, new ParameterizedTypeReference() {
            }).getBody();
            LOG.log(Level.INFO, "/elk trying to print exception - &gt; " + exceptionrsp);
            response = response + " === " + exceptionrsp;
        } catch (Exception e) {
            // exception should not reach here. Really bad practice
        }
        return response;
    }

    @RequestMapping(value = "/exception")
    public String exception() {
        String rsp = "";
        try {
            int i = 1 / 0;
            // should get exception
        } catch (Exception e) {
            e.printStackTrace();
            LOG.error(e);
            StringWriter sw = new StringWriter();
            PrintWriter pw = new PrintWriter(sw);
            e.printStackTrace(pw);
            String sStackTrace = sw.toString(); // stack trace as a string
            LOG.error("Exception As String :: - &gt; "+sStackTrace);
            rsp = sStackTrace;
        }
        return rsp;
    }
}
```

打开资源文件夹下的 application.properties 并添加以下配置。

```
logging.file=elk-example.log
spring.application.name = elk-example
```

启动应用程序,通过浏览 http://localhost:8080/elk 进行测试。转到应用程序根目录,并验证是否已创建日志文件(即 elk-example.log)。两次访问端点,并验证日志已添加到日志文件中。

创建一个 logstash 配置文件,以便它侦听日志文件并将日志消息推送到 Elasticsearch 中。这是示例中使用的 logstash 配置,请根据实际的设置更改日志路径,代码如下:

```
input {
  file {
    type => "java"
    path => "F:/Study/eclipse_workspace_mars/elk-example-spring-boot/elk-example.log"
    codec => multiline {
      pattern => "^%{YEAR}-%{MONTHNUM}-%{MONTHDAY} %{TIME}.*"
      negate => "true"
      what => "previous"
    }
  }
}

filter {
  #If log line contains tab character followed by 'at' then we will tag that entry as stacktrace
  if [message] =~ "\tat" {
    grok {
      match => ["message", "^(\tat)"]
      add_tag => ["stacktrace"]
    }
  }

  grok {
    match => [ "message", "(?<timestamp>%{YEAR}-%{MONTHNUM}-%{MONTHDAY} %{TIME}) %{LOGLEVEL:level} %{NUMBER:pid} --- \[(?<thread>[A-Za-z0-9-]+)\] [A-Za-z0-9.]*\.(?<class>[A-Za-z0-9#_]+)\s*:\s+(?<logmessage>.*)",
               "message", "(?<timestamp>%{YEAR}-%{MONTHNUM}-%{MONTHDAY} %{TIME}) %{LOGLEVEL:level} %{NUMBER:pid} --- .+? :\s+(?<logmessage>.*)"
             ]
  }

  date {
    match => [ "timestamp" , "yyyy-MM-dd HH:mm:ss.SSS" ]
  }
}

output {
  stdout {
    codec => rubydebug
  }
  # Sending properly parsed log events to elasticsearch
  elasticsearch {
    hosts => ["localhost:9200"]
```

```
    }
}
```

在 Kibana 中查看日志前，需要配置索引模式。在这里，可以将 logstash-* 配置为默认配置。虽然可以在 logstash 端更改此索引模式，并在 Kibana 中进行配置，但为简单起见，我们将使用默认配置。

7.3 本章小结

本章介绍了使用 Spring Boot Framework 的子项目 Spring Boot Actuator 来监控和管理 Spring Boot 应用程序，还介绍了使用 Elastic 栈实现应用程序日志监控。

参考文献

[1] SEDGEWICK R, WAYNE K. 算法[M]. 谢路云, 译. 4 版. 北京：人民邮电出版社. 2012.

[2] HORSTMANN C S. Java 核心技术·卷 I：基础知识[M]. 林琪, 苏钰涵, 等译. 11 版. 北京：机械工业出版社，2019.

[3] 算法[EB/OL]. http://algs4.cs.princeton.edu/home/, 2019-01-01.